T0320797

Visible Light Communication

Visible light communication (VLC) is an evolving communication technology for short-range applications. Exploiting recent advances in the development of high-power visible-light-emitting LEDs, VLC offers an energy-efficient, clean alternative to RF technology, enabling the development of optical wireless communication systems that make use of existing lighting infrastructure.

Drawing on the expertise of leading researchers from across the world, this concise book sets out the theoretical principles of VLC, and outlines key applications of this cutting-edge technology. Providing insight into modulation techniques, positioning and communication, synchronization, and industry standards, as well as techniques for improving network performance, this is an invaluable resource for graduate students and researchers in the fields of visible light communication and optical wireless communication, and for industrial practitioners in the field of telecommunications.

Shlomi Arnon is a Professor at the Department of Electrical and Computer Engineering at Ben-Gurion University (BGU), Israel. He is a Fellow of SPIE, a co-editor of *Advanced Optical Wireless Communication Systems* (2012), and has edited special issues of the *Journal of Optical Communications and Networking* (2006) and the *IEEE Journal on Selected Areas in Communications* (2009, 2015).

Visible Light Communication

EDITED BY

SHLOMI ARNON

Ben-Gurion University of the Negev, Israel

CAMBRIDGE
UNIVERSITY PRESS

CAMBRIDGE
UNIVERSITY PRESS

University Printing House, Cambridge CB2 8BS, United Kingdom

Cambridge University Press is part of the University of Cambridge.

It furthers the University's mission by disseminating knowledge in the pursuit of education, learning and research at the highest international levels of excellence.

www.cambridge.org
Information on this title: www.cambridge.org/9781107061552

© Cambridge University Press 2015

This publication is in copyright. Subject to statutory exception and to the provisions of relevant collective licencing agreements, no reproduction of any part may take place without the written permission of Cambridge University Press.

First published 2015

A catalog record for this publication is available from the British Library

Library of Congress Cataloging in Publication data
Visible light communication / edited by Shlomi Arnon, Ben Gurion University of the Negev, Israel.
    pages   cm
ISBN 978-1-107-06155-2 (hardback)
1. Optical communications.   I. Arnon, Shlomi, 1968–
TK5103.59.V57   2015
621.382'7–dc23

                                          2014046702

ISBN 978-1-107-06155-2 Hardback

Cambridge University Press has no responsibility for the persistence or accuracy of URLs for external or third-party internet websites referred to in this publication, and does not guarantee that any content on such websites is, or will remain, accurate or appropriate.

Every effort has been made in preparing this book to provide accurate and up-to-date information which is in accord with accepted standards and practice at the time of publication. Nevertheless, the authors, editors and publishers can make no warranties that the information, including, but not limited to, any methods, formulae, and instructions, contained herein is totally free from error. The authors, editors and publishers therefore disclaim all liability for direct or consequential damages resulting from the use of material contained in this book. Readers are strongly advised to pay careful attention to information provided by the manufacturer of any equipment that they plan to use.

Contents

9   Image sensor based visible light communication   181

Shinichiro Haruyama and Takaya Yamazato

Contributors

Shlomi Arnon
Ben-Gurion University of the Negev, Israel

Bo Bai
Northwestern Polytechnical University, China

Chen Gong
University of Science and Technology of China

Shinichiro Haruyama
Keio University, Japan

Mohsen Kavehrad
The Pennsylvania State University, USA

Jae Kyun Kwon
Yeungnam University, Korea

Klaus-Dieter Langer
Fraunhofer Heinrich Hertz Institute (HHI), Germany

Sang Hyun Lee
Sejong University, Korea

Kang Tae-Gyu
Electronics and Telecommunications Research Institute
(ETRI), Korea

Zixiong Wang
The Hong Kong Polytechnic University

Zhengyuan Xu
University of Science and Technology of China

Takaya Yamazato
Nagoya University, Japan

Weizhi Zhang
The Pennsylvania State University, USA

Wen-De Zhong
Nanyang Technological University (NTU), Singapore

Acknowledgment

I would like to express my special appreciation and thanks to my beloved wife, who has helped me for so many years, and for the interesting discussions with my kids that always challenge me. I also want to thank all the book contributors, who have made it what it is – without them the book would never have been realized.

1   Introduction

Shlomi Arnon

Visible light communications (VLC) is the name given to an optical wireless communication system that carries information by modulating light in the visible spectrum (400–700 nm) that is principally used for illumination [1–3]. The communications signal is encoded on top of the illumination light. Interest in VLC has grown rapidly with the growth of high power light emitting diodes (LEDs) in the visible spectrum. The motivation to use the illumination light for communication is to save energy by exploiting the illumination to carry information and, at the same time, to use technology that is "green" in comparison to radio frequency (RF) technology, while using the existing infrastructure of the lighting system. The necessity to develop an additional wireless communication technology is the result of the almost exponential growth in the demand for high-speed wireless connectivity. Emerging applications that use VLC include: a) indoor communication where it augments WiFi and cellular wireless communications [4] which follow the smart city concept [5]; b) communication wireless links for the internet of things (IOT) [6]; c) communication systems as part of intelligent transport systems (ITS) [7–14]; d) wireless communication systems in hospitals [15–17]; e) toys and theme park entertainment [18, 19]; and f) provision of dynamic advertising information through a smart phone camera [20].

VLC to augment WiFi and cellular wireless communication in indoor applications has become a necessity, with the result that many people carry more than one wireless device at any time, for example a smart phone, tablet, smart watch, and smart glasses and a wearable computer, and at the same time the required data rate from each device is growing exponentially. It is also becoming increasingly clear that in urban surroundings, human beings spend most of their time indoors, so the practicality of VLC technology is self-evident. It would be extremely easy to add extra capacity to existing infrastructure by installing a VLC system in offices or residential premises. In Fig. 1.1 we can see an example of a VLC network that provides wireless communication to a laptop, smart phone, TV, and wearable computer.

The downlink includes illumination LED, Ethernet power line communication (PLC) modem, and LED driver, which receives a signal by a dedicated or dongle receiver as part of the device. The uplink configuration could be based for example on: a) a WiFi link; b) an infrared–IRDA link; or c) a modulated retro reflector (Fig. 1.2). A modulated retro reflector is an optical device that retro reflects incident light [21, 22]. The amplitude of the retro reflected light is controlled by an electronic signal, as a result modulation of the light can be achieved. In the cases of an infrared–IRDA [23] link or a modulated retro reflector

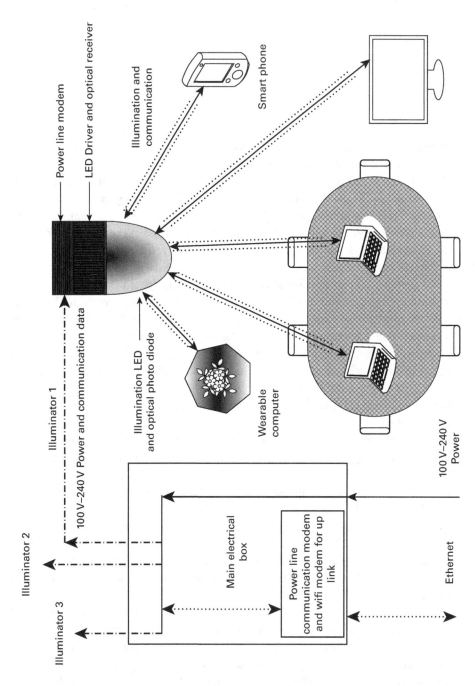

Figure 1.1 VLC wireless network.

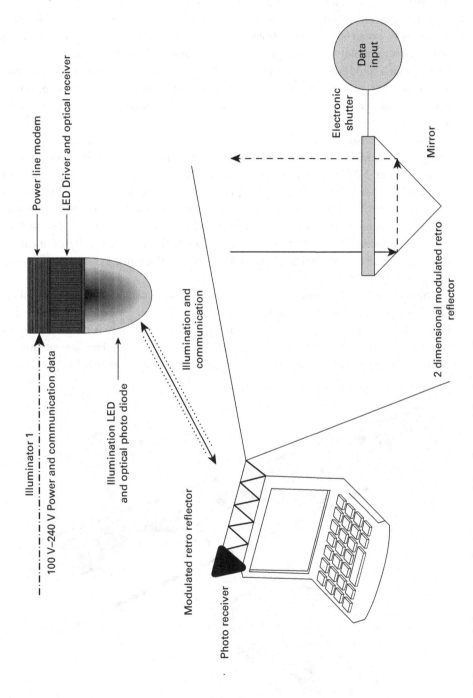

Figure 1.2 A wireless communication network based on modulated retro reflector.

the receiver could be part of the illumination LED. In that case the uplink receiver includes photo diode, trans impedance amplifier and modem. In this way, an operational wireless network could be created in next to no time.

In the near future, billions of appliances, sensors and instruments will have wireless connectivity, as can be anticipated from the revolutionary concept of the internet of things (IOT). This technology makes it possible to have ambient intelligence and autonomous control which could adapt the environment to the requirements and the desires of people. VLC could be a very relevant wireless communication technology that is cheap, simple and immediate and does not encroach on an already crowded part of the electromagnetic spectrum.

Intelligent transport systems (ITS) are an emerging technology for increasing road safety and reducing the number of road casualties as well as for improving traffic efficiency (Fig. 1.3). VLC have been proposed as a means for providing inter-vehicular communication and for establishing connectivity between vehicular and road infrastructure, such as traffic lights and billboards. These systems provide one-way or two-way short- to medium-range wireless communication links that are specifically designed for the automotive sphere. The technology uses the headlights and the rear lights of cars as transmitters, and cameras and detectors as the receivers. The traffic lights are the counterpart of a transmitter in this sphere.

The medical community pursues ways to improve the efficiency of hospitals and at the same time to reduce hospital-acquired infections, which are very costly in money and human life. One way to upgrade the communication infrastructure is by wireless technology. The technology makes it possible for doctors to access and update patient data using tablet computers at the patient's bedside instead of manually keeping paper

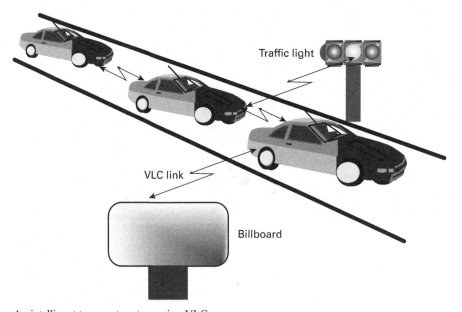

Figure 1.3   An intelligent transport system using VLC.

documents that are kept either at the bedside or in the nurses' back office station. Another application is a device used to monitor patient well-being and vital data remotely. Due to the fact that RF, like WiFi and cellular, is a best effort technology, meaning that the transmission of information is not guaranteed, interference from nearby devices could jam the communication. This situation is unacceptable for medical applications and therefore a switch to VLC technology is self-evident. It is clear that VLC technology can provide localized solutions immune to interference and jamming for the medical domain.

The toys and theme park entertainment sector is a very interesting application that takes advantage of VLC technology due to two main characteristics (Fig. 1.4). The first characteristic is the ability to communicate by line of sight or semi line of sight, so the communication is localized to a specific volume. This makes it possible to provide location based information in the theme park so the audience will have a multi-dimensional and multi-sensory experience. The same concept could be used in the toys market to communicate between toys, using the already present LED. The second characteristic is the low cost required to implement the technology in toy and park

Figure 1.4   VLC in toys in the theme park entertainment industry (courtesy of S. Arnon).

entertainment. One example of a way to reduce the cost of toys is using the toy LED simultaneously as transmitter and photo diode receiver.

Dynamic advertising through a smart phone camera is a new area that uses a billboard and illumination to transmit information that is detected by the camera, and then by an appropriate algorithm the communications data are extracted from the video. This technology could be used to add an extra and dynamic layer of information to advertisements in the street, shopping center and subway.

The IEEE has defined a new standard, IEEE 802.15.7, which describes high data rate VLC, up to 96 Mb/s, by fast modulation of optical light sources. Meanwhile, in several experiments around the world data rates of more than 500 Mb/s have been reported [24–27]. In addition many new methods are being developed to maximize bit rates and to provide algorithms to manage interference and subcarrier re-use [27–33], methods to analyze synchronization issues [34] and methods to design an asymmetric communication system using modulated retro-reflectors (MRR) based on nanotechnology [35]. In the light of these developments a new methodological book covering this subject is required in order to help progress the technology. The aspiration of this book is to serve as a textbook for undergraduate and graduate level courses for students in electrical engineering, electro-optical engineering, communications engineering, illumination engineering and physics. It is also intended to serve as a source for self-study and as a reference book for senior engineers involved in the design of VLC systems.

The book includes nine chapters that cover the important aspects of VLC scientific theory and technology. Following this introductory chapter, Chapter 2 deals with modulation techniques under lighting constraints, and is written by Jae Kyun Kwon and Sang Hyun Lee. The authors explain that the physical layer design of VLC systems is of substantially different character than conventional wireless systems, notably with regard to a new average intensity constraint. This new constraint, which will be referred to as the "dimming target" of the illumination system, introduces a new domain of system design that has rarely been considered in existing communication media. Chapter 3, by Wen-De Zhong and Zixiong Wang, describes methods such as receiver plane tilting and a special LED lamp arrangement as performance enhancement techniques for indoor VLC systems. The next two chapters, 4 and 5, deal with very important aspects of VLC, which take advantage of the characteristics of light to obtain location information. Chapter 4, by Mohsen Kavehrad and Weizhi Zhang, outlines the applications, investigates current access to the radio spectrum and studies in depth the shifting needs of indoor positioning in the visible light spectrum. At the end of the chapter, challenges and potential solutions are discussed. Chapter 5, by Zhengyuan Xu, Chen Gong, and Bo Bai, presents indoor and outdoor light positioning systems (LPS). For indoor LPS, combining VLC with position estimation methods is presented and an optimal estimation algorithm is implemented at the receiver to provide an unbiased estimate of the camera position. For outdoor automotive LPS, the light signal could be emitted from a traffic light, carrying its position information. Then, the position of the vehicle could be estimated based on the received position information of the traffic light and the time difference of arrival (TDoA) of the light signal at two photodiodes. Chapter 6, by Kang Tae-Gyu describes the standards for VLC. In this chapter, the lamp and electronic power are presented in terms of electric safety in IEC TC 34. Later, other standards for VLC, such as PLASA E1.45 and IEEE 802.15.7, which need a number of protocols between a sending party and a receiving party are discussed, as well as electric safety. This chapter also discusses the compatibilities of the VLC service area, illumination, vendor considerations and standards. Chapter 7, by Shlomi Arnon, presents different modulation methods used in VLC, such as on off keying (OOK) pulse position modulation (PPM), inverse pulse position modulation (IPPM) and variable pulse position modulation (VPPM). Later, explanations are given on how to calculate the bit error rate (BER) for each modulation scheme. A detailed description of how to calculate the effect of synchronization time offset and clock jitter on the BER performance is also presented. Chapter 8, by Klaus-Dieter Langer, describes discrete multitone (DMT) modulation for VLC and presents advanced and highly spectrally efficient solutions for this scheme, such as DC-biased DMT, asymmetrically clipped optical OFDM (ACO-OFDM) and pulse-amplitude-modulated discrete multitone (PAM-DMT). Chapter 9, by Shinichiro Haruyama and Takaya Yamazato, deals with image sensor based VLC and introduces two unique applications using an image sensor: (1) massively parallel visible light transmissions that can theoretically achieve a maximum data rate of 1.28 Gigabits per second; and (2) accurate sensor position estimation that cannot be realized by a VLC system using a single-element photodiode (PD). Applications of image sensor based communication for the automotive industry, position measurements in civil engineering and bridge position monitoring are also presented.

References

[1] Shlomi Arnon, John Barry, George Karagiannidis, Robert Schober, and Murat Uysal, eds., *Advanced Optical Wireless Communication Systems*. Cambridge University Press, 2012.

[2] Sridhar Rajagopal, Richard D. Roberts, and Sang-Kyu Lim, "IEEE 802.15.7 visible light communication: Modulation schemes dimming support." *Communications Magazine, IEEE* **50**, (*3*), 2012, 72–82.

[3] IEEE Standard 802.15.7 for local and metropolitan area networks – Part 15.7: Short-range wireless optical communication using visible light, 2011.

[4] Cheng-Xiang Wang, Fourat Haider, Xiqi Gao, *et al.*, "Cellular architecture and key technologies for 5G wireless communication networks." *IEEE Communications Magazine* **52**, (*2*), 2014, 122–130.

[5] Shahid Ayub, Sharadha Kariyawasam, Mahsa Honary, and Bahram Honary, "A practical approach of VLC architecture for smart city." In *Antennas and Propagation* Conference (LAPC), 2013 Loughborough, 106–111, IEEE, 2013.

[6] Tetsuya Yokotani, "Application and technical issues on Internet of Things." In *Optical Internet (COIN)*, 2012 10th International Conference, 67–68, IEEE, 2012.

[7] Fred E. Schubert, and Jong Kyu Kim, "Solid-state light sources getting smart." *Science* **308**, (*5726*), 2005, 1274–1278.

[8] Shlomi Arnon, "Optimised optical wireless car-to-traffic-light communication." *Transactions on Emerging Telecommunications Technologies* **25**, 2014, 660–665.

[9] Seok Ju Lee, Jae Kyun Kwon, Sung-Yoon Jung, and Young-Hoon Kwon, "Evaluation of visible light communication channel delay profiles for automotive applications." *EURASIP Journal on Wireless Communications and Networking (1)*, 2012, 1–8.

[10] Sang-Yub Lee, Jae-Kyu Lee, Duck-Keun Park, and Sang-Hyun Park, "Development of automotive multimedia system using visible light communications." In *Multimedia and Ubiquitous Engineering*, pp. 219–225. Springer, 2014.

[11] S.-H. Yu, Oliver Shih, H.-M. Tsai, and R. D. Roberts, "Smart automotive lighting for vehicle safety." *Communications Magazine, IEEE* **51**, (*12*), 2013, 50–59.

[12] Shun-Hsiang You, Shih-Hao Chang, Hao-Min Lin, and Hsin-Mu Tsai, "Visible light communications for scooter safety." In Proceeding of the 11th Annual International Conference on *Mobile Systems, Applications, and Services*, 509–510, ACM, 2013.

[13] Zabih Ghassemlooy, Wasiu Popoola, and Sujan Rajbhandari, *Optical Wireless Communications: System and Channel Modelling with Matlab®*. CRC Press, 2012.

[14] Alin Cailean, Barthelemy Cagneau, Luc Chassagne, *et al.*, "Visible light communications: Application to cooperation between vehicles and road infrastructures." In *Intelligent Vehicles Symposium (IV)*, 1055–1059, IEEE, 2012.

[15] Ryosuke Murai, Tatsuo Sakai, Hajime Kawano, *et al.*, "A novel visible light communication system for enhanced control of autonomous delivery robots in a hospital." In *System Integration (SII)*, 2012 IEEE/SICE International Symposium, 510–516, IEEE, 2012.

[16] Seyed Sina Torkestani, Nicolas Barbot, Stephanie Sahuguede, Anne Julien-Vergonjanne, and J.-P. Cances, "Performance and transmission power bound analysis for optical wireless based mobile healthcare applications." In *Personal Indoor and Mobile Radio Communications (PIMRC)*, 22nd International Symposium, 2198–2202, IEEE, 2011.

[17] Yee Yong Tan, Sang-Joong Jung, and Wan-Young Chung, "Real time biomedical signal transmission of mixed ECG signal and patient information using visible light communication." In *Engineering in Medicine and Biology Society (EMBC)*, 35th Annual International Conference of the IEEE, 4791–4794, IEEE, 2013.

[18] Nils Ole Tippenhauer, Domenico Giustiniano, and Stefan Mangold, "Toys communicating with leds: Enabling toy cars interaction." In *Consumer Communications and Networking Conference (CCNC)*, 48–49, IEEE, 2012.

[19] Stefan Schmid, Giorgio Corbellini, Stefan Mangold, and Thomas R. Gross, "LED-to-LED visible light communication networks." In Proceedings of the fourteenth ACM international symposium on *Mobile ad hoc Networking and Computing*, 1–10, ACM, 2013.

[20] Richard D. Roberts, "Undersampled frequency shift ON-OFF keying (UFSOOK) for camera communications (CamCom)." In *Wireless and Optical Communication* Conference (WOCC), 645–648, IEEE, 2013.

[21] Etai Rosenkrantz, and Shlomi Arnon, "Modulating light by metal nanospheres-embedded PZT thin-film." *Nanotechnology, IEEE Transactions on* **13**, (*2*), 222–227, March 2014.

[22] Etai Rosenkrantz and Shlomi Arnon, "An innovative modulating retro-reflector for free-space optical communication." In SPIE Optical Engineering + Applications, p. 88740D. International Society for Optics and Photonics, 2013.

[23] Rob Otte, *Low-Power Wireless Infrared Communications*. Springer-Verlag, 2010.

[24] Yuanquan Wang, Yiguang Wang, Chi Nan, Yu Jianjun, and Shang Huiliang, "Demonstration of 575-Mb/s downlink and 225-Mb/s uplink bi-directional SCM-WDM visible light communication using RGB LED and phosphor-based LED." *Optics Express* **21**, (*1*), 2013, 1203–1208.

[25] Ahmad Helmi Azhar, T. Tran, and Dominic O'Brien, "A gigabit/s indoor wireless transmission using MIMO-OFDM visible-light communications." *Photonics Technology Letters, IEEE* **25**, (*2*), 2013, 171–174.

[26] Wen-Yi Lin, Chia-Yi Chen, Hai-Han Lu, *et al.*, "10m/500Mbps WDM visible light communication systems." *Optics Express* **20**, (*9*), 2012, 9919–9924.

[27] Fang-Ming Wu, Chun-Ting Lin, Chia-Chien Wei, *et al.*, "3.22-Gb/s WDM visible light communication of a single RGB LED employing carrier-less amplitude and phase modulation." In *Optical Fiber Communication* Conference, OTh1G-4. Optical Society of America, 2013.

[28] Giulio Cossu, Raffaele Corsini, Amir M. Khalid, and Ernesto Ciaramella, "Bi-directional 400 Mbit/s LED-based optical wireless communication for non-directed line of sight transmission." In *Optical Fiber Communication* Conference, p. Th1F–2. Optical Society of America, 2014.

[29] Dima Bykhovsky and Shlomi Arnon, "Multiple access resource allocation in visible light communication systems." *Journal of Lightwave Technology* **32**, (*8*), 2014, 1594–1600.

[30] Dima Bykhovsky and Shlomi Arnon, "An experimental comparison of different bit-and-power-allocation algorithms for DCO-OFDM." *Journal of Lightwave Technology* **32**, (*8*), 2014, 1559–1564.

[31] Joon-ho Choi, Eun-byeol Cho, Zabih Ghassemlooy, Soeun Kim, and Chung Ghiu Lee, "Visible light communications employing PPM and PWM formats for simultaneous data transmission and dimming." *Optical and Quantum Electronics* 1–14, May 2014.

[32] Nan Chi, Yuanquan Wang, Yiguang Wang, Xingxing Huang, and Xiaoyuan Lu, "Ultra-high-speed single red-green-blue light-emitting diode-based visible light communication system utilizing advanced modulation formats." *Chinese Optics Letters* **12**, (*1*), 2014, 010605.

[33] Liane Grobe, Anagnostis Paraskevopoulos, Jonas Hilt, *et al.*, "High-speed visible light communication systems." *Communications Magazine, IEEE* **51**, (*12*), 2013, 60–66.

[34] Shlomi Arnon, "The effect of clock jitter in visible light communication applications." *Journal of Lightwave Technology* **30**, (*21*), 2012, 3434–3439.

[35] Etai Rosencrantz and Shlomi Arnon, "Tunable electro-optic filter based on metal-ferroelectric nanocomposite for VLC," *Optics Letters* **39**, (*16*), 2014, 4954–4957.

2 Modulation techniques with lighting constraints

Jae Kyun Kwon and Sang Hyun Lee

The physical layer design of visible light communication (VLC) systems is of substantially different characteristics from standard RF communications in that it involves a new constraint, namely a lighting constraint. This kind of constraint is imposed on the average intensity and flicker of the light emission. Since the flicker has little impact on human eye perception when light pulses blink at 200 Hz or higher frequency, the average intensity constraint is mainly addressed in this chapter. While this constraint is usually given as an inequality in optical wireless communication, it is represented as an equality in VLC. In addition, compared to radio-frequency communication, where the signal power, the squared value of signal level, is usually constrained, the intensity, the signal level itself, is constrained. In other words, the lighting constraint is defined with respect to the average (the first-order moment) of the signal, instead of the variance (the second-order moment). Therefore, this new constraint, which will be referred to as the dimming target, introduces a new domain of system design which has rarely been considered in existing communication media.

In this chapter, several ways of communicating a message subject to the average constraint are addressed. This chapter is far from comprehensive but attempts to offer several promising ways of achieving such a goal. To satisfy the lighting constraint represented by the average constraint, several approaches have been addressed, and they can be categorized as to shift signal levels, to compensate in time, and to change level distribution. Some of those schemes are simply realized, and some provide improved throughput. First, the shift of the signal level is one of the simplest approaches. A typical non-return-to-zero (NRZ) on-off keying (OOK) has 50% average intensity for uniform probability of binary symbols. For the lighting constraint of 75% dimming, it offers a simple solution of moving the OFF symbol level from 0% intensity to 50%, which is referred to as analog dimming. Although this is conceptually simple, a non-linear characteristic of LEDs poses some technical difficulties, and the reduced level spacing degrades detection performance.

Second, compensation of the intensity difference in time is another approach that can be simply realized. For general data transmission with 50% average intensity, i.e., uniform symbol probability, when the lighting constraint of 75% dimming is targeted, the same duration of *dummy* ON transmission time as the data transmission time is appended to meet the target. If the target is below 50%, a dummy OFF symbol interval is applied to resolve the difference. Those dummy transmissions may be either appended after each

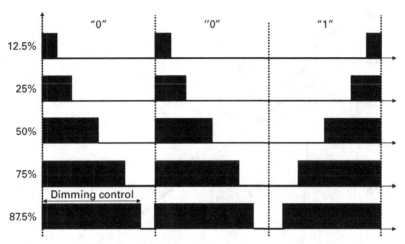

Figure 2.1   PPM with a dimming control (after [9]).

single data frame or inserted between every symbol. An example of the former is the time-multiplexed OOK [2, 3] presented in the IEEE standard. On the other hand, pulse width modulation (PWM) is a typical representation of the latter. Several studies [4, 5, 6] based on PWM present some simple solutions to provide a marginal rate enhancement. PWM can also be superposed with OOK and pulse position modulation (PPM) for dimming support [7, 8]. Variable pulse position modulation (VPPM) [2] is another approach that uses PWM. This combines 2-PPM and PWM for a dimming control as shown in Fig. 2.1. Pulse dual slope modulation [10] is a variant of VPPM and offers improved flicker mitigation.

Third, change of the distribution of symbol levels can be considered a sophisticated approach, which promises additional rate enhancement. Inverse source coding (ISC) [11, 12] converts the uniform distribution of ON and OFF symbol levels to 75% ON symbols and 25% OFF symbols for achieving a binary OOK with 75% target. This can also be extended to M-ary modulation and provides a theoretical data rate bound asymptotically in a noise-free environment. Multiple-PPM (MPPM) [3, 13, 14] is another scheme that falls in this category. This uses all possible combinations of ON and OFF symbols within a specified interval to represent distinct messages, while the ratio of ON symbols and OFF symbols is adjusted to meet the dimming target. Although ISC and MPPM provide high throughput in a low-noise channel, both still have some challenges to overcome to coexist with channel coding in a high-noise channel. To this end, several practical schemes [15–18] to accommodate channel coding in dimmable VLC have been proposed.

Typical LED lighting provides white illumination. However, some applications require multi-colored LEDs, such as light therapy, display, and lighting with a high color rendering index. In such applications, the lighting requirement does not simply give a constraint on scalar average intensity but a vector of average color and intensity. Two approaches for the colored case are addressed here. Color shift keying (CSK) [2, 19, 20]

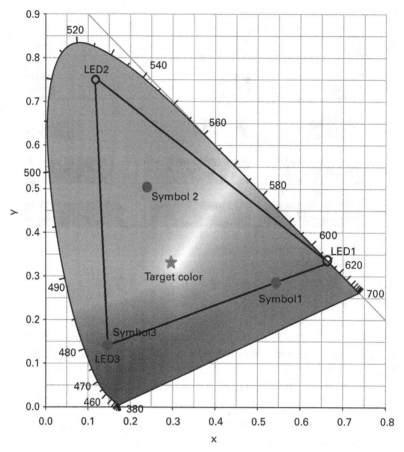

Figure 2.2   Symbol constellation of CSK with three symbols in CIE xyY color space (© 2012 IEEE, reprinted, with permission, from [21]).

separates consideration of color and intensity. The intensity is fixed to the target intensity, and the data transmission is carried out using an instantaneous variation of the color, as shown in Fig. 2.2. The signal constellation lies in a two-dimensional color space with the identical intensity. On the other hand, color intensity modulation (CIM) [21] changes both the color and intensity simultaneously. Therefore, this can provide throughput enhancement over CSK.

In the following sections, the detail of three schemes is considered: ISC, multi-level approaches, and CIM. In addition, another physical consideration should be taken into account in the implementation of the system. VLC requires instantaneous variation of lighting to avoid human detection. Thus, when the variation of symbols is fast enough this ensures no flicker. In addition, careful attention should also be given to physical color temperature and chromaticity shift of LED lighting. This results mainly from the change in input current level and temperature of the LED, and multi-level schemes are susceptible to this shift.

2.1   Inverse source coding in dimmable VLC

2.1.1   ISC for NRZ-OOK

The ISC scheme for binary modulation is introduced first. Let d denote the dimming target. To achieve this target, the OOK modulation should be formed using ON and OFF symbols in the proportion of d and $1 - d$ respectively. If the communication is carried out using this modulation, the data rate is upper-bounded by the binary entropy, which is given by

$$E_p = -d \log_2 d - (1 - d) \log_2 (1 - d). \tag{2.1}$$

Thus, to achieve the maximum transmission efficiency (data rate) with the dimming target d, the composition of message symbols should be adjusted so that ON and OFF symbols occur with probability d and $1 - d$, respectively, in a single data frame. Since the source coding (compression) operation is used to maximize the entropy by changing the composition of symbols as uniformly as possible, the inverse operation of this can be applied to adjust the composition of symbols to an arbitrary proportion. Thus, this operation is referred to as inverse source coding (or dimming coding) and can be incorporated in the transmitter of a VLC system as in Fig. 2.3. Since the proportion of input binary symbols is kept to be even, a binary scrambling operation may be applied to maintain the uniform input probability in the case that the symbol composition of the input message is non-uniform. The binary scrambling can be realized by taking symbol-wise modulo-2 operations after adding a random binary sequence to a stream of the message symbol.

Figure 2.4 shows the transmission efficiency increase of ISC over existing time-multiplexing-based dimming support. The transmission efficiency of the ISC, denoted by E_p, is consistently improved over the existing scheme, denoted by E_0, and both become equal when $d = 0$, 0.5, and 1. Also, the increase in efficiency $\frac{E_p}{E_0}$ is expressed as

$$\frac{E_p}{E_0} = \frac{-d \log_2 d - (1 - d) \log_2 (1 - d)}{2d}, \tag{2.2}$$

as illustrated in Fig. 2.4. As the dimming target deviates from 0.5, the efficiency improvement becomes larger. A 50% improvement in efficiency is induced when the dimming target is 29% (or 71%) and 100% for 16% (or 84%).

Here, the realization of ISC is considered and exemplified using a Huffman code. Let the dimming target d be 0.7. That is, the proportion of ON and OFF symbols should be 70% and 30%, respectively. Thus, Huffman encoding for this condition is first applied. Since the probability of an ON symbol is larger than that of an OFF symbol, the ON symbol is jointly considered with the consecutive symbols. Thus, the resulting probabilities for new symbols "0," "10," and "11" are given as in Table 2.1. The rightmost column in Table 2.1 lists codewords that are encoded by the Huffman code. The average lengths of uncoded and encoded symbols are 1.7 and 1.51, respectively. Thus, the compression ratio is given as $\frac{1.51}{1.7} \approx 0.888$. Since the entropy can be evaluated as $-0.3 \log_2 0.3 - 0.7 \log_2 0.7 \approx 0.881$, more than 94% $\left(< \frac{1-0.888}{1-0.881} \right)$ of the compression ratio is obtained compared to the maximally achievable compression ratio. Inverse

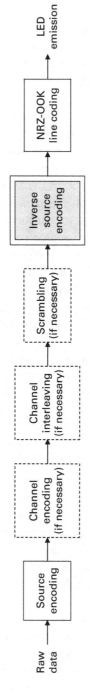

Figure 2.3 The transmitter of a VLC system with inverse source coding (© 2010 IEEE, reprinted, with permission, from [11]).

Table 2.1 Huffman encoding (© 2010 IEEE. Reprinted, with permission, from [11]).

Symbol / length	Probability	Codeword / length
0 / 1	0.3	00 / 2
10 / 2	0.21	01 / 2
11 / 2	0.49	1 / 1

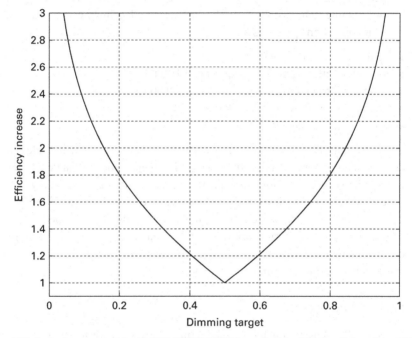

Figure 2.4   Efficiency improvement with ISC (© 2010 IEEE, reprinted, with permission, from [11]).

Huffman coding is used to convert a data stream with uniform binary symbols to that with 70% of the resulting symbols being ON. This is shown in Table 2.2, which is realized simply via the inverse of the mapping presented in Table 2.1. The average lengths of uncoded and inversely-Huffman-encoded symbols are 1.5 and 1.75, respectively. Thus, the decompression ratio is $\frac{1.75}{1.5} \approx 1.17 \approx \frac{1}{0.857}$. The resulting dimming rate is given by

$$\frac{0 \times \frac{1}{4} + \left(1 \times \frac{1}{4} + 0 \times \frac{1}{4}\right) + \left(1 \times \frac{1}{2} + 1 \times \frac{1}{2}\right)}{1 \times \frac{1}{4} + 2 \times \frac{1}{4} + 2 \times \frac{1}{2}} = \frac{1.25}{1.75} = 0.714, \qquad (2.3)$$

which is close to the dimming target of $d = 0.7$. Elaborate Huffman coding and its associated inverse Huffman coding using a larger number of symbols can yield an improved fit to the dimming target.

Table 2.2 Inverse Huffman encoding (© 2010 IEEE. Reprinted, with permission, from [11]).

Symbol / length	Probability	Codeword / length
00 / 2	0.25	0 / 1
01 / 2	0.25	10 / 2
1 / 1	0.5	11 / 2

Lastly, the conflict between channel coding and inverse source coding is addressed. Let an example sequence 00 01 1 01 1 be inversely Huffman-coded. Then, the flow of encoding operations proceeds as:

- input sequence: 00 01 1 01 1,
- inversely-Huffman-encoded sequence: 0 10 11 10 11,
- sequence corrupted by erroneous channel: 0 00 11 10 11,
- sequence aligned for recovery: 0 0 0 11 10 11,
- recovered sequence: 00 00 00 1 01 1.

In this example, the number of symbols increases by 2 and the fourth single symbol "1" is decoded to three-tuple symbol "000." Thus, the channel coding that may follow is highly like to fail to recover the original sequence. This indicates that an ordinary inverse source code may not seem to work well with the usual channel coding. Therefore, for VLC transmission over erroneous channels, two topics for ISC remain open: the existence of an ISC approach which can work well with channel coding, and the design of a channel coding scheme that can produce codewords of unequal binary probability of ON and OFF symbols that adapt to the dimming target.

2.1.2   ISC for M-ary PAM

In ISC with OOK modulation, the dimming target straightforwardly determines the binary symbol probability by adjusting the duty cycle of a data frame, the proportion of ON and OFF symbols. However, non-binary modulation, such as pulse amplitude modulation, may employ various alternatives for dimming support, each of which leads to a different value of the spectral efficiency. In this context, the distribution of non-binary symbols that maximizes the spectral efficiency is considered. For the spectral efficiency, the entropy can be considered. The entropy for M-PAM is simply given by

$$-\sum_{i=1}^{M} p_i \log_2 p_i, \qquad (2.4)$$

where p_i is the probability of the ith level of PAM. In the equidistant M-PAM, the spacing between two adjacent levels is equal and the dimming target d is expressed as

$$d = \sum_{i=1}^{M} \frac{i-1}{M-1} p_i, \qquad (2.5)$$

since the ith level lies on $\frac{i-1}{M-1}$ of the maximum level and d is the normalized average of levels. The symbol probability distribution $\{p_i\}$ that maximizes (2.4) under the constraint

(2.5) is obtained via an optimization formulation. Since this optimization is concave, its solution is guaranteed to provide the global maximum. To obtain a closed-form solution of the optimization, the dual formulation [1] is derived as

$$\mathcal{L}(\{p_i\}, \lambda_1, \lambda_2) = -\sum_{i=1}^{M} p_i \log_2 p_i - \lambda_1 \sum_{i=1}^{M} p_i - \lambda_2 A \sum_{i=1}^{M} \frac{i-1}{M-1} p_i, \qquad (2.6)$$

where λ_1 and λ_2 are Lagrange multipliers. Then, by some algebra, the dimming target can be expressed as

$$d = \frac{2^{-a}}{(1-r)(M-1)} \left(\frac{r(1-r^{M-1})}{1-r} - (M-1)r^M \right), \qquad (2.7)$$

where $a = 1/\ln 2 + \lambda_1$ and $r = 2^{-\frac{\lambda_2 A}{M-1}}$. Thus, a feasible pair of (λ_1, λ_2) is chosen such that the distribution $\{p_i\}$ is well-defined and maximizes the entropy. To implement the obtained distribution, the inverse Huffman code may be used. Figure 2.5 shows the normalized entropy of 3,4,8-PAM ISC and time-multiplex dimming. ISC consistently outperforms the time-multiplex dimming scheme regardless of the modulation order. For M-PAM ISC, the normalized entropies are all of similar trend regardless of M. Therefore, ISC remains efficient for any large M.

2.1.3   Comparisons with respect to dimming capacity

Here, ISC, analog dimming and their hybrid approach are compared. The analog dimming approach changes the intensity level of symbols: the intensity level is raised if the

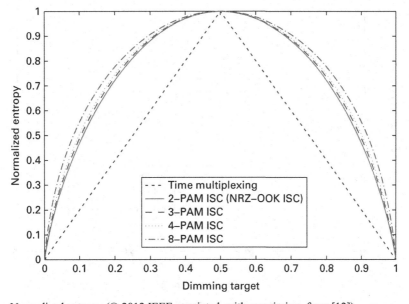

Figure 2.5   Normalized entropy (© 2012 IEEE, reprinted, with permission, from [12]).

dimming target is greater than 0.5, and reduced if it is smaller than 0.5. The intensity lies within the interval $[0, A]$ and the symbol spacing is equidistant. The largest shift in symbol intensity is $2A(d - 0.5)$. Hybrid dimming is achieved for an intensity shift in analog dimming smaller than $2A|d - 0.5|$ and ISC is used to adjust the remaining shift towards the dimming target. The entropy of analog dimming remains constant for the dimming target. Since the entropy curve of 3-PAM hybrid dimming results from the horizontal scaling of 3-PAM ISC, 3-PAM hybrid dimming has the same maximum entropy as that of 50% dimming for an arbitrary dimming target. Note that other PAMs such as 3-PAM and 6-PAM are used in this chapter for explanatory purposes, though 2^n-PAMs are practically used. Since 95% dimming is expected to yield a lower data rate than 50% dimming in a noisy channel, special care is required for consideration of the performance degradation caused by noise in comparison with analog or hybrid dimming. The minimum distance between symbols is considered here. If the minimum distance is identical, the order of performance degradation can be assumed to be equal. Thus, the entropy of the dimming methods can be compared when the minimum distance between symbols is identical. Figure 2.6 depicts the entropy of 4-PAM ISC and 3-PAM hybrid dimming with the same minimum distance. 4-PAM ISC is consistently superior to 3-PAM hybrid dimming. This is straightforward because the symbol set of 3-PAM hybrid dimming $\{S_2, S_3, S_4\}$ can be considered a special case where the probability of S_1 is zero in the symbol set of 4-PAM ISC, $\{S_1, S_2, S_3, S_4\}$. Therefore, M-PAM ISC is better than or equal to N-PAM hybrid dimming in performance when $M > N$ with the same minimum distance. However, the comparison of ISC, analog dimming, and hybrid dimming is not

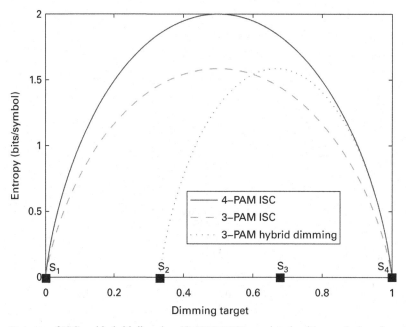

Figure 2.6   Entropy of ISC and hybrid dimming (© 2012 IEEE, reprinted, with permission, from [12]).

straightforward for different minimum distances. Thus, the dimming capacity is introduced to compare between dimming methods for different minimum distances.

Dimming strategies are compared under additive white Gaussian noise (AWGN):

$$Y = X + Z,\tag{2.8}$$

where X is a transmitted signal, Y is a received signal, and Z is AWGN with zero mean and variance σ^2. The dimming capacity, denoted by $C_d \equiv I(X;Y)$, is defined as the mutual information between X and Y subject to the dimming constraint $\frac{E[X]}{A} = d$. Then, $I(X; Y)$ is given by

$$I(X; Y) = -\int_{-\infty}^{\infty} f_Y(y)\log_2 f_Y(y)dy - \frac{1}{2}\log_2(2\pi e\sigma^2),\tag{2.9}$$

where $f.(\cdot)$ is the probability distribution function, so that

$$f_X(x) = \sum_{i=1}^{M} p_i\delta(x - b_i),\tag{2.10}$$

$$f_Y(y) = \sum_{i=1}^{M} p_i f_Z(y - b_i),\tag{2.11}$$

where

$$b_i = \begin{cases} \dfrac{(A - D_s)(i - 1)}{M - 1} + D_s & \text{if } d \geq 0.5 \\ \dfrac{(A - D_s)(i - 1)}{M - 1}, & \text{otherwise,} \end{cases}\tag{2.12}$$

and D_s is the level shift for analog and hybrid dimming. Thus, the intensity range $[0, A]$ can change into either $[0, A - D_s]$ or $[D_s, A]$ according to the dimming target.

Figure 2.7 shows the dimming capacity of ISC by 2,3,4,8-PAM for dimming target 0.5. The best modulation order that achieves the maximum value of the capacity changes with respect to channel quality measure A/σ. Since the average power cannot be adjusted freely under a given dimming target, analog dimming can reduce the intensity range of $[0, A]$ only. Thus, the decrease in dimming capacity by analog dimming is directly indicated in Fig. 2.7 by a leftward shift along the horizontal axis, A/σ. For example, 3-PAM is used and dimming changes from 0.5 to 0.8. The intensity range changes from $[0, A]$ to $[0.6A, A]$ in order to satisfy 0.8 analog dimming, where the range is limited to only 40% of the original range. This results in a horizontal shift of -3.98 dB $\simeq -4.0$ dB. If A/σ is 9 dB, this becomes equivalently 5 dB for analog dimming. Then, the dimming capacity of 0.8 analog dimming corresponds to 0.69, decreased from 1.47 in the two filled squares shown in Fig. 2.7.

Here, the performance comparison among ISC, analog dimming, and hybrid dimming is discussed. The modulation level is restricted to 2,3,4,8, and 16-PAM. Figure 2.8 depicts the dimming capacity with respect to A/σ and dimming target. Figure 2.9

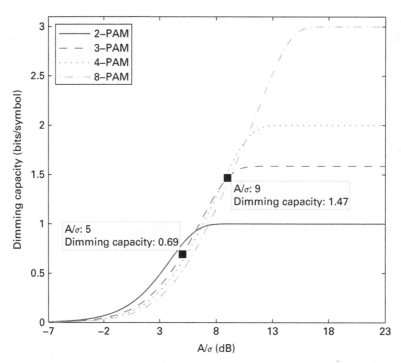

Figure 2.7   The dimming capacity of *M*-PAM with 0.5 dimming (© 2012 IEEE, reprinted, with permission, from [12]).

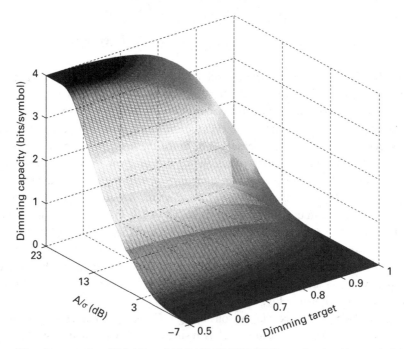

Figure 2.8   Dimming capacity of 2,3,4,8, and 16-PAM (© 2012 IEEE, reprinted, with permission, from [12]).

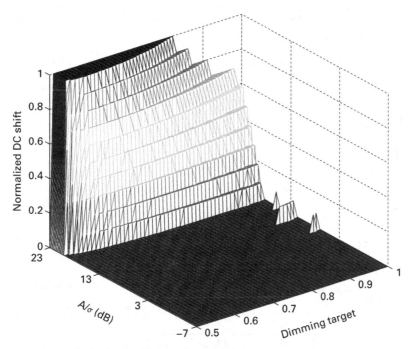

Figure 2.9   Dimming method selection (© 2012 IEEE, reprinted, with permission, from [12]).

illustrates the dimming method that yields the maximum capacity. Figure 2.9 exhibits the best dimming method with respect to A/σ and the dimming target. ISC and analog dimming are chosen when the normalized DC shift (intensity shift) is zero and one, respectively. Hybrid dimming can be applied for mid-range values. ISC has better performance except when A/σ is greater than 20 dB and when A/σ is around 11.5 dB with the dimming target 97%. If higher modulations are allowed for a region A/σ greater than 20 dB, ISC still remains the best dimming method. Another exception occurs around an A/σ of 11.5 dB, where modulations between 4-PAM and 8-PAM are not allowed. Figure 2.10 shows how to determine the y-axis coordinate in Fig. 2.9 when A/σ is 11.5 dB and the dimming target is 97%. The maximum dimming capacity occurs for 6-PAM and ISC. If 6-PAM is not allowed, 4-PAM with slight DC shift (hybrid dimming) yields the maximum. However, the difference of the capacity is negligible between ISC and hybrid dimming in 4-PAM in this case. Therefore, ISC is the best method when all modulation levels are available, and ISC is of performance better than or similar to the others when only a subset of modulation levels is allowed.

2.2   Multi-level transmission in dimmable VLC

In this section, a multi-level transmission scheme is presented for visible light communication systems of dimming support. To provide a dimming control of the multi-level modulation, the concatenation of different pulse-amplitude-modulated symbols is used

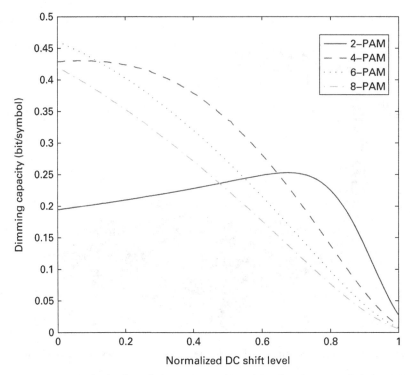

Figure 2.10   Dimming method selection when A/σ is 11.5 dB and the dimming target is 97% (© 2012 IEEE, reprinted, with permission, from [12]).

to yield the overall signal with an average amplitude that matches the dimming requirement. This scheme also aims at arbitrary dimming adaptation by adjusting the composition of concatenated differently-modulated symbols. To this end, the problem is formulated into linear programming with the objective of maximizing the transmission data rate and satisfying the dimming requirements.

The need for the improved spectral efficiency leads to the use of multi-level modulations, such as PAM, and poses a new challenge for the multi-level transmission scheme design that adapts to the dimming requirement.

To this end, this section presents a practical multi-level transmission approach that uses a concatenated code where each of the component codes is encoded and constructed with different modulations. To obtain high error-correcting capability with a simple code structure, a group of linear codes is used and concatenated. However, a linear code can only generate a set of codewords with a uniform number of symbols. The average intensity level of a linear codeword modulated using PAM of the peak level intensity A is always equal to $\frac{A}{2}$, which corresponds to the dimming of 0.5. In other words, the straightforward concatenation of linear component codes does not allow for the adaptation to non-trivial dimming requirements. Therefore, the design of an efficient transmission scheme satisfying the desired dimming requirement is pursued by adjusting the proportion of symbols linearly encoded and modulated with different

PAMs and concatenating those symbols. Since the proportion of linear codes is adjusted arbitrarily, the adaptation to an arbitrary dimming requirement can be accomplished. In what follows, a linear optimization formulation that determines the optimal composition of linear coded symbols for the highest spectral efficiency is introduced and solved.

2.2.1   Multi-level transmission scheme

A model multi-level transmission scheme is first introduced and a linear optimization is formulated for obtaining the best configuration of the scheme. For this scheme, N message symbols that constitute a single data frame are either modulated in one of $M-1$ different PAMs (2 to M-PAM) or punctured. The spacings between levels are uniform for all modulations. That is, the difference between any two adjacent levels of a PAM is identical. Without loss of generality, the dimming target d is within $[0,0.5]$, i.e., $d \in [0,0.5]$. An example of such a multi-level transmission scheme is depicted in Fig. 2.11. The horizontal and vertical axes are associated with the symbol configuration (or the symbol order) and the intensity of the transmitted signal, respectively. Message symbols modulated in different modulations are serially concatenated. Thus, a message symbol modulated in one of the M PAMs is transmitted at each symbol time instant. However, the message symbols are not necessarily transmitted in this order, because the transmission of the symbols in this way causes a periodic gradual decrease of the intensity in a data frame, which results in a severe flickering effect. Therefore, the symbols are interleaved in a random way known a priori to transmitter and receiver so that the order of the intensity levels is randomized and the flicker is averaged out. A level of one PAM is aligned with the same level of different PAMs. For example, the ith level of j-PAM is of the same intensity as the ith level of k-PAM for $i \leq j < k$. The unit level of the vertical axis in Fig. 2.11 is uniform. Thus, the transmission scheme can be thought of as an M-ary PAM with non-uniform symbol level probability or

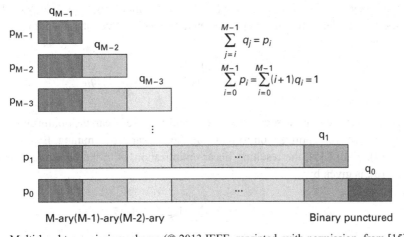

Figure 2.11   Multi-level transmission scheme (© 2013 IEEE, reprinted, with permission, from [15]).

alternatively as the superposition of $M-1$ different OOK modulations. The values of effective power are different for the messages modulated in different PAMs since the mean symbol power increases with increasing number of levels of PAM. If the message is assumed to be random and modulated in $(i+1)$-PAM, each level in a PAM symbol occurs with uniform probability $1/(i+1)$ and the average symbol level is equal to $i/2$ times the intensity of a single level. By changing the proportion of symbols modulated in M different PAMs, the dimming rate can be adjusted to an arbitrary target. The proportions of level $i=0,\ldots,M-1$ and that of each level in $(i+1)$-alphabet symbols are denoted by p_i and q_i, respectively. It is defined that $q_{M-1}=p_{M-1}$. Then, it follows that

$$q_i = \begin{cases} p_i - p_{i+1} & \text{if } i = 0,\ldots,M-2, \\ p_{M-1} & \text{if } i = M-1. \end{cases} \qquad (2.13)$$

Note that $p_i = \sum_{j=i}^{M-1} q_j$ and the sum of $\{p_i\}$ is equal to one. Also, it is obvious that

$$\sum_{i=0}^{M-1} p_i = \sum_{i=0}^{M-1}\sum_{j=i}^{M-1} q_j = \sum_{i=0}^{M-1}(i+1)q_i = 1. \qquad (2.14)$$

That is, the $(i+1)$-PAM has the proportion of $(i+1)q_i$, which corresponds to the proportion of the ith column in Fig. 2.11. A symbol modulated in $(i+1)$-PAM has the amount of information proportional to $\log_2(i+1)$ bits per symbol. Since the dimming target is associated with the average intensity, distribution $\{p_i\}$ can be chosen to meet the dimming target. Therefore, the dimming target is expressed with respect to distribution $\{p_i\}$ as

$$d = \sum_{i=0}^{M-1} \frac{i}{M-1} p_i. \qquad (2.15)$$

When $d \in (0.5,1]$, a distribution $\{\bar{p}_i\}$ can be redefined as $\bar{p}_i = p_{M-i-1}$ by symmetry. For this purpose, the proportion for $(i+1)$-PAM q_i is found such that the overall spectral efficiency is maximized. Since the $(i+1)$-PAM symbol can convey $\log_2(i+1)$ bits, the spectral efficiency (or entropy) is chosen to be an objective function of the optimization, which is given by

$$E(\{q_i\}) \equiv \sum_{i=0}^{M-1}((i+1)\log_2(i+1))q_i. \qquad (2.16)$$

Since all expressions associated with $\{p_i\}$ and $\{q_i\}$ are linear, the optimization is given in a linear convex optimization formulation, i.e., linear programming. By merging (2.13), (2.14), and (2.15), the resulting linear optimization, with the objective of maximizing (2.16), is given by

$$\max_{\bar{q}_i \geq 0} \sum_{i=0}^{M-1}(\log_2(i+1))\bar{q}_i$$

subject to

$$\sum_{i=0}^{M-1} \overline{q}_i = 1,$$

$$\sum_{i=0}^{M-1} i\overline{q}_i = 2(M-1)d,$$

$$\overline{q}_i = \begin{cases} (i+1)(p_i - p_{i+1}) & \text{if } i = 0, \ldots, M-2, \\ Mp_{M-1} & \text{if } i = M-1. \end{cases} \tag{2.17}$$

To obtain a closed-form solution, the dual formulation [1] is derived because only two (dual) variables are sufficient and the solution is easier to obtain than the original (primal) formulation. Therefore, (2.17) is reformulated as:

$$\min_{\lambda,\mu} \quad \lambda + 2(M-1)d\mu$$

subject to

$$\lambda + i\mu \geq \log_2(i+1), \ i = 0,\ldots,M-1. \tag{2.18}$$

To see this, the Lagrange function [1] of (2.17) can be considered. Since there are two constraints, two Lagrange multipliers denoted by λ and μ, respectively, are introduced. Since the Lagrange function provides the original formulation with a concave upper bound, the dual formulation turns out to be the minimization of the bound in order to match the bound to the maximum. For this goal, the Lagrange function can be expanded with respect to M variables $\overline{q}_i$. Then, the fact that the coefficients of those M variables become negative in the minimization leads to the corresponding M different constraints, and the remaining terms that are independent of the variables provide the objective function of (2.18). The objective function depends on symbol alphabet M and dimming target d. Also, each constraint is associated with each of M PAMs. Since both formulations are simple linear programming, the strong duality holds by Slater's condition [1]. In other words, the optimal solutions of both formulations are identical. The dual formulation can also be interpreted in a geometric way as illustrated in Fig. 2.12. The M different constraints generate a convex feasible region over the (λ,μ) plane. In addition, the objective function is associated with a line across the feasible region with its slope being $1/(2(M-1)d)$ and its x-intercept equal to the objective value. Therefore, the dual problem turns out to determine a line which touches a point of the feasible region and has the smallest x-intercept. According to the constraints in (2.18), there are M different lines of decreasing slopes that define the boundary of the feasible region in the plane. Therefore, $M-1$ intersecting points can occur on the boundary of the feasible region. Since the feasible region is convex, the line associated with the objective function intersects only a single point of the feasible region. By inspection, the solution occurs at a point where only two constraints in (2.18) hold the equality. Therefore, consideration of the intersection of those points and the line associated with the objective function

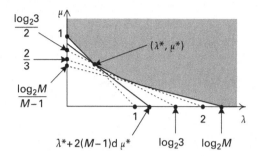

Figure 2.12   Geometric interpretation of the dual problem (© 2013 IEEE, reprinted, with permission, from [15]).

suffices. By the complementary slackness, all variables $\bar{q}_i$ associated with strict inequalities have zero values, i.e., $\bar{q}_i = 0$ for i such that $\lambda + i\mu > \log_2(i + 1)$. At most, two primal variables $\bar{q}_i$ can take non-zero values, and only two (consecutive) PAMs are needed for the optimal transmission. Here, the effective modulation parameter is defined as $\bar{m} \equiv \lfloor 2(M - 1)d \rfloor$. Then, the solution is expressed as

$$\lambda^* = \log_2 \frac{(\bar{m} + 1)^{\bar{m}+1}}{(\bar{m} + 2)^{\bar{m}}}, \quad \mu^* = \log_2 \frac{\bar{m} + 2}{\bar{m} + 1}. \tag{2.19}$$

The corresponding objective value is given by

$$E(\lambda^*, \mu^*) = \log_2(\bar{m} + 1)\left(\frac{\bar{m} + 2}{\bar{m} + 1}\right)^{2(M-1)d-\bar{m}}. \tag{2.20}$$

In addition, the associated distribution is obtained as

$$\bar{q}_i^* = \begin{cases} 1 - \left(2(M - 1)d - \bar{m}\right) & \text{if } i = \bar{m}, \\ 2(M - 1)d - \bar{m} & \text{if } i = \bar{m} + 1, \\ 0 & \text{otherwise.} \end{cases} \tag{2.21}$$

This implies that only two modulations of $(\bar{m} + 1)$-PAM and $(\bar{m} + 2)$-PAM are used in the optimal transmission. This can be considered in a general setting where an arbitrary subset of PAM alphabets is allowed instead of all alphabets (1 to M-PAM). Since a single constraint in (2.18) is associated with a distinct modulation, the dual formulation and its geometric interpretation are intact for any change in PAMs. Since the number of message symbols is practically chosen to be power-of-twos, 2^k-PAMs are preferred for positive integer k. For example, only 1, 2, 4, and 8-PAM are used. Then, the corresponding feasible region becomes a polygon with at most four vertices. The intersections and the corresponding data rates are listed along with valid ranges of the dimming rate in Table 2.3. The upper triangular part (i.e. above the table diagonal) shows the intersections associated with two PAMs corresponding to row and column indices. The lower triangular part represents the corresponding objective value. The last row and column correspond to the lower and upper limits of achievable dimming targets. For example, in the case of only 2-PAM and 4-PAM used, the line associated with the objective

Table 2.3 Solution for the system consisting of 1,2,4, and 8-PAM (© 2013 IEEE, reprinted, with permission, from [15]).

(PAM,PAM)	1	2	4	8	B
1		(0,1)	$\left(0,\dfrac{2}{3}\right)$	$\left(0,\dfrac{3}{7}\right)$	0
2	$14d$		$\left(\dfrac{1}{2},\dfrac{1}{2}\right)$	$\left(\dfrac{2}{3},\dfrac{1}{3}\right)$	$\dfrac{1}{14}$
4	$\dfrac{28d}{3}$	$\dfrac{1+14d}{2}$		$\left(\dfrac{5}{4},\dfrac{1}{4}\right)$	$\dfrac{3}{14}$
8	$6d$	$\dfrac{2+14d}{3}$	$\dfrac{5+14d}{4}$		$\dfrac{1}{2}$
A	0	$\dfrac{1}{14}$	$\dfrac{3}{14}$	$\dfrac{1}{2}$	$d \in [A,B]$

function crosses point $\left(\frac{1}{2},\frac{1}{2}\right)$ on the (λ, μ) plane. The corresponding objective value is $\frac{1+14d}{2}$ for $d \in \left[\frac{1}{14},\frac{3}{14}\right]$.

To ensure that all levels of the $(i + 1)$-PAM symbol occur uniformly, $(i + 1)$-ary scrambling can be applied by taking symbol-wise modulo-$(i + 1)$ operations after addition of an $(i + 1)$-ary random sequence. For example, 4-PAM is used and transmitted symbols are assumed to be 0 1 1 2 2 2 3 1. Thus, the symbol level probability is non-uniform. In addition, the scrambling sequence is given as 3 2 1 0 3 2 1 0. The symbols obtained after scrambling are given as 3 3 2 2 5 4 4 1 mod 4 = 3 3 2 2 1 0 0 1. Therefore, the resulting transmitted symbols are uniform. To retrieve the original symbols at the receiver, the scrambling sequence is subtracted from the received symbols and the recovered symbols are given as 0112 –2 –2 –11 mod 4 = 0 1 1 2 2 2 3 1. Let N_i be the number of symbols modulated in $(i + 1)$-PAM in a data frame. That is, N_i is the closest integer to $N\bar{q}_i$. Then, the overall dimming rate is considered as a random variable D such that $E[D] = d$ and is given by

$$D = \frac{\displaystyle\sum_{i=0}^{M-1}\sum_{j=0}^{N_i-1}\frac{A}{M-1}U_{ij}}{\displaystyle\sum_{i=0}^{M-1}\sum_{j=0}^{N_i-1}A}, \tag{2.22}$$

where U_{ij} is a discrete random variable uniformly distributed over interval $[0, i]$, representing the jth symbol modulated in $(i + 1)$-PAM. Also, A is the largest level of M-PAM and the spacing between two adjacent levels is equal to $\frac{A}{M-1}$. Therefore, the variance of D denoted by var$[D]$ is given by

$$\text{var}[D] = \frac{1}{N(M-1)^2}\sum_{i=0}^{M-1}\frac{i(i+2)}{12}\bar{q}_i. \tag{2.23}$$

Since only $(\overline{m} + 1)$-PAM and $(\overline{m} + 2)$-PAM are used for the optimal transmission, the variance is simply given by

$$\text{var}[D] = \frac{(\overline{m} + 1)(\overline{m} + 3) - (2\overline{m} + 3)\overline{q}_{\overline{m}}^*}{12N(M - 1)^2} \leq \frac{(M + 1)}{12N(M - 1)}, \qquad (2.24)$$

where the equality holds when $d = 0.5$. Since the lowest optical clock rate is 200 kHz [22], the variation can be bounded within approximately 1% by taking the data frame length N greater than 1000 for $M = 8$. Therefore, flicker does not occur because the brightness of visible light changes at a rate much faster than 150–200 Hz [2].

2.2.2   Asymptotic performance

The performance of the multi-level transmission scheme is addressed in various configurations, such as all consecutive alphabets and power-of-two alphabets. For fair comparison, the normalized capacity is evaluated by dividing the average number of bits per symbol by $\log_2(i + 1)$ for $(i + 1)$-PAM, as illustrated in Fig. 2.13. This corresponds to the relative data rate introduced in order to compare various transmission schemes with different data rates. ML-$(i + 1)$PAM denotes the multi-level transmission scheme that allows consecutive $(i + 1)$ different PAMs (1 to $(i + 1)$-PAM), and rML-4PAM(1,4) denotes a transmission scheme with only 1-PAM and 4-PAM used. Although ISC [12] establishes a theoretical upper bound, a practical capacity-approaching scheme achieving

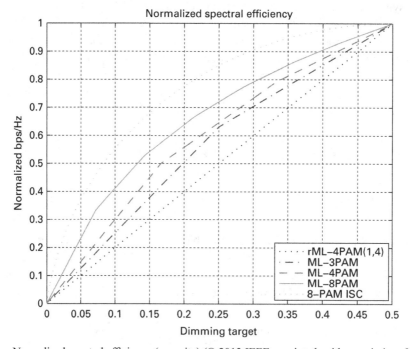

Figure 2.13   Normalized spectral efficiency (capacity) (© 2013 IEEE, reprinted, with permission, from [15]).

this bound has not been found. The spectral efficiency of M-PAM ISC increases slowly as the number of alphabets M increases. The performance of the scheme approaches the upper bound in a piecewise linear way for the increasing number of alphabets. Thus, a good upper-bound for ML-MPAM in Fig. 2.13 is obtained. The kth point where the slope changes is located at $\left(\frac{k}{2(M-1)} , \frac{\log_2(k+1)}{\log_2 M} \right)$ on the ML-MPAM curve. As the number of alphabet M increases, the number of such points increases correspondingly, and the set of line segments connecting adjacent points provides a good approximation of the curve for ML-MPAM. Then, all points in $\left\{ \left(\frac{k}{2(M-1)} , \frac{\log_2(k+1)}{\log_2 M} \right) | k = 0, \ldots , M - 1 \right\}$ lie along the curve represented by $f_M(x) = \frac{\log_2(1+2(M-1)x)}{\log_2 M}$. Note that for $x \in (0, 0.5]$, it becomes the case that $f_M(x) = 1$ as M tends to infinity.

The performance improvement over rML-$(i + 1)$PAM$(1, i + 1)$ is depicted in Fig. 2.14. The improvement decreases gradually as the dimming rate approaches 0.5. The optimal composition of PAMs is shown for various dimming rates in Fig. 2.15. This is consistent with the solution of the optimization where at most two PAMs are chosen for the optimal transmission. Since eight different PAMs are allowed (1 to 8-PAM), two lines associated with k-PAM and $(k + 1)$-PAM cross in interval $\left[\frac{k-1}{14} , \frac{k}{14} \right]$ for $k = 1, \ldots, 7$, i.e., at seven different points. Compared to this, three crosses occur in rML-$8PAM(1,2,4,8)$. Therefore, this figure is used to determine the optimal

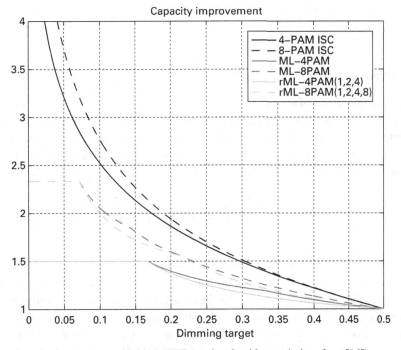

Figure 2.14   Capacity improvement (© 2013 IEEE, reprinted, with permission, from [15]).

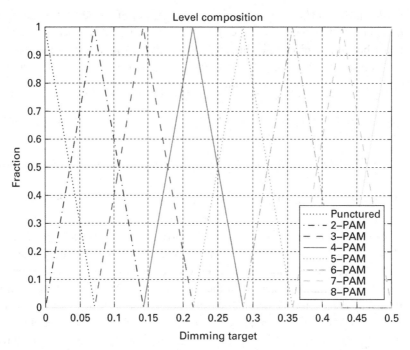

Figure 2.15   Level composition (© 2013 IEEE, reprinted, with permission, from [15]).

transmission scheme graphically. Once the target dimming rate is known a priori, both the transmitter and receiver can find the optimal composition of the modulations immediately.

2.2.3   Simulation results

Simulation results of uncoded and coded transmission schemes are shown here. Let A and M be the largest level and the order of the largest ordered PAM, respectively, i.e., A is the intensity of the Mth level of M-PAM. Under the assumption of uniform PAM symbol level, the kth level of $(i+1)$-PAM is expressed as $\frac{kA}{M-1}$. Then, the symbol error probability of the uncoded $(i+1)$-PAM is given by

$$P_{\text{err}}^{(i+1)} = \frac{2i}{i+1} Q\left(\frac{1}{2(M-1)}\frac{A}{\sigma}\right) = \frac{2i}{i+1} Q_M\left(\frac{A}{\sigma}\right), \qquad (2.25)$$

where $Q_M(x) \equiv Q\left(\frac{x}{2(M-1)}\right)$, and σ^2 is Gaussian noise power. For a channel quality measure, the signal-intensity-to-noise-amplitude ratio is used. In optical wireless communication, the LED modulates the instantaneous optical intensity proportional to an input electrical current signal via intensity modulation and direct detection [23]. The combination of these two conversion steps ensures the validity of (2.25) in VLC. The overall symbol error probability is given by

$$\overline{P}_{\text{err}} = \sum_{i \in \mathcal{M}} \overline{q}_i \frac{2i}{i+1} \, \mathcal{Q}_M\left(\frac{A}{\sigma}\right), \tag{2.26}$$

where $\mathcal{M}$ is the set of the largest levels (in $[0, M-1]$) of all PAMs allowed in a data frame, i.e., $\mathcal{M} = \{i \in [0, M-1] | \overline{q}_i > 0\}$. Since only two adjacent PAMs are used for the optimal scheme in (2.21), the error probability is given by

$$\overline{P}_{\text{err}}^* = 2 \frac{(\overline{m}+1)^2 - \overline{q}_{\overline{m}}^*}{(\overline{m}+1)(\overline{m}+2)} \, \mathcal{Q}_M\left(\frac{A}{\sigma}\right) \leq 2 \frac{\overline{m}+1}{\overline{m}+2} \, \mathcal{Q}_M\left(\frac{A}{\sigma}\right). \tag{2.27}$$

The decoding error performance depends on two parameters $\overline{m}$ and $\overline{q}_{\overline{m}}^*$, and hence on dimming rate d and modulation order M. The equality holds when $\overline{q}_{\overline{m}}^* = 0$, i.e., $d = \frac{\overline{m}+1}{2(M-1)}$ for $\overline{m} = 0, \ldots, M-2$ and, when $d = 0$, no information is transmitted and the error bound obviously becomes zero. The error bound is maximized if only M-PAM is used and $d = 0.5$. This is obvious because the symbol error probability inherently grows as the number of transmitted symbols increases. Given that $P_{\text{err}}^{(i+1)} > \varepsilon$ for a non-zero ε, spectral efficiency $R^{(i+1)}$ decreases to zero as fast as $O((1 - \varepsilon \overline{q}_i)^N)$. Therefore, it is necessary to use a good channel code for each PAM to obtain high spectral efficiency. The resulting spectral efficiency is expressed as

$$\overline{R}^* = \overline{q}_{\overline{m}}^* \left(1 - 2 \frac{\overline{m}}{\overline{m}+1} \, \mathcal{Q}_M\left(\frac{A}{\sigma}\right)\right)^{N \overline{q}_{\overline{m}}^*} \log_2(\overline{m}+1)$$

$$+ (1 - \overline{q}_{\overline{m}}^*) \left(1 - 2 \frac{\overline{m}+1}{\overline{m}+2} \, \mathcal{Q}_M\left(\frac{A}{\sigma}\right)\right)^{N(1-\overline{q}_{\overline{m}}^*)} \log_2(\overline{m}+2). \tag{2.28}$$

Since $\overline{m}$ and $\overline{q}_{\overline{m}}^*$ are functions of the dimming target, so is the spectral efficiency.

The performance of coded schemes is now considered. Symbol error probability $P_{\text{err}}^{(i+1)}$ for $i \in \mathcal{M}$ is obtained only by simulation. If the information is encoded with code rate R and modulated into N PAM symbols, the two exponents of (2.28) are replaced with $NR\overline{q}_{\overline{m}}^*$ and $NR(1 - \overline{q}_{\overline{m}}^*)$, respectively. Turbo codes are used for the evaluation of the coded performance, in that the use of practical coding schemes guarantees the feasibility of the transmission scheme, and turbo codes lead to the capacity-approaching performance. In addition, turbo codes are robust against the performance degradation incurred by puncturing for adapting an arbitrary dimming target. Figures 2.16 and 2.17 compare the spectral efficiencies for dimming requirement $d = 0.1$ and 0.4, respectively. Several different codes of rates $R = \frac{1}{3}, \frac{1}{2}, \frac{3}{4}$ can be obtained from a single code of rate $\frac{1}{3}$ by puncturing. For comparison, the results of other transmission schemes that meet the dimming requirement, using 8-PAM or $(\overline{m}+2)$-PAM are presented. For the uncoded scheme, the result of a random scheme is presented by calculating the average with respect to the ensemble of feasible profiles $\{\overline{q}_i\}$. For different dimming targets, the multi-level transmission scheme consistently shows superior throughput performance to others, and it selects different pairs of PAMs accordingly, e.g., (2,3)-PAM for $d = 0.1$, (3,4)-PAM for $d = 0.2$, (5,6)-PAM for $d = 0.3$, and (6,7)-PAM for $d = 0.4$. Between two

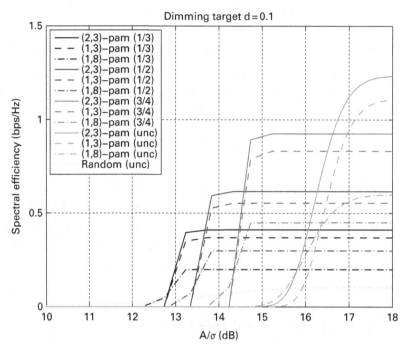

Figure 2.16   Spectral efficiency for dimming target 0.1 (© 2013 IEEE, reprinted, with permission, from [15]).

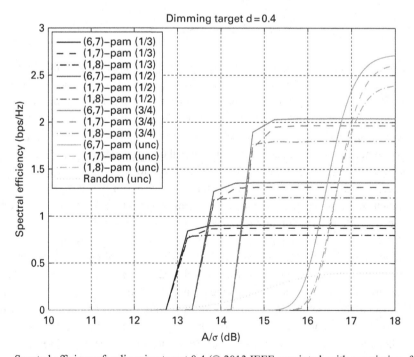

Figure 2.17   Spectral efficiency for dimming target 0.4 (© 2013 IEEE, reprinted, with permission, from [15]).

cases that use puncturing, the scheme using $(\overline{m} + 2)$-PAM outperforms that using the maximal order PAM (8-PAM). The performance difference between them is large in a low dimming regime, because larger numbers of symbols are punctured with smaller values of d for 8-PAM and the overall amount of information is lower than for $(\overline{m} + 2)$-PAM.

2.3   Color intensity modulation for multi-colored VLC

2.3.1   Color space and signal space

Now the characteristics of the color space are introduced and the difference from the signal space is considered. To this end, the model of the multi-colored system is briefly described. N different LEDs and corresponding photodetectors (PDs) with different wavelength characteristics are used. Let $I_i(\lambda)$ and $r_j(\lambda)$ denote the intensity of the ith LED and the responsivity of the jth PD, respectively. Then, the overall intensity at the receivers is given by $I(\lambda) = \sum_{i=1}^{N} I_i(\lambda)$. Let $\overline{x}(\lambda)$, $\overline{y}(\lambda)$, and $\overline{z}(\lambda)$ be the normalized color matching functions [24] associated with the color perception capability of human eyes. Then, three components of light stimuli are given by $X = \int \overline{x}(\lambda)I(\lambda)d\lambda$, $Y = \int \overline{y}(\lambda)I(\lambda)d\lambda$, and $Z = \int \overline{z}(\lambda)I(\lambda)d\lambda$. These parameters can characterize the perception in human eyes indicated in CIE XYZ color space [24]. A single color is associated with each line passing the origin in CIE XYZ color space, as shown in Fig. 2.18, and its corresponding intensity is represented by the distance of a point on the line from the origin. To provide a VLC feature under these color matching and dimming constraints, color shift keying (CSK) [22] can be used. CSK places the symbol constellation around the target color in CIE xyY color space such that the average of colors associated with symbols is the same as the target color. At the same time, the intensity of LED output is controlled to meet the dimming requirement. Figure 2.2 shows an example of CSK achieved using three LEDs, denoted by LED1, LED2, and LED3, respectively. Message symbols can be placed within a triangle with three vertices associated with those three LEDs such that the overall average coincides with the point corresponding to the target color. In doing so, message symbols should be located as far as possible from one another to minimize the detection error. The detection of the signal is carried out in the signal space. Thus, the received signal is represented with an N-dimensional vector $\mathbf{S} = [S_1, \ldots, S_N]^T$, where the ith component, the output of the ith photodetector, is $S_i = \int r_i(\lambda)I(\lambda)d\lambda$. Lighting conditions of color matching and dimming are considered in the color space, while the spectral efficiency (or mutual information) of the communication feature can be addressed in the signal space. Therefore, it is not straightforward to consider both objectives simultaneously in a single space.

2.3.2   Color intensity modulation

To mitigate this difficulty and improve the spectral efficiency under the given constraint, color intensity modulation (CIM) is proposed. To this end, a subspace (or a

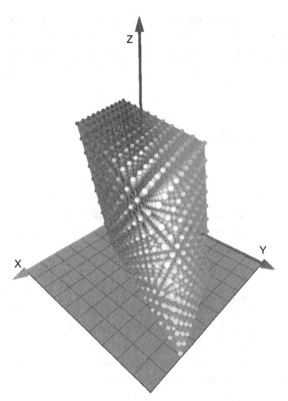

Figure 2.18   CIE XYZ color space.

point) in the signal space is chosen which is associated with the subspace of the color space where color constraint is satisfied. For this subspace of the signal space, the positions of symbols are determined to maximize the spectral efficiency subject to the constraint that the symbols' weighted average belongs to the subspace. For the control of the average intensity of symbols, PAM is used here since it has the same bandwidth regardless of the average intensity. When the average intensity is achieved using M-PAM ISC [12], two different approaches can be used: either adjusting symbol probabilities for fixed equidistant symbols or controlling positions and probabilities simultaneously. In Figs. 2.19 and 2.20, those two approaches are illustrated with A/σ 8 dB and the dimming target 0.8. The second approach offers 1.3% improved performance with the simultaneous control of positions and probabilities of message symbols.

Now multi-dimensional channel is considered with multi-color LEDs. For independent and parallel control of optical channels, wavelength division multiplexing (WDM) can be used. ISC offers each optical channel of WDM with color matching and dimming control. On the other hand, CIM does not suffer from inter-channel interference, even when the channels are non-orthogonal. CIM yields improved spectral efficiency over the normal WDM and WDM with ISC in non-orthogonal channels. To summarize the

Figure 2.19   Optimal equidistant symbols with mutual information 0.9373 bits/symbol when A/σ is 8 dB, and the dimming target is 80% (© 2012 IEEE, reprinted, with permission, from [21]).

Figure 2.20   Optimal adjusted symbols with mutual information 0.9494 bits/symbol when A/σ is 8 dB, and the dimming target is 80% (© 2012 IEEE, reprinted, with permission, from [21]).

concepts of CSK, WDM, and CIM, first, CSK uses only a two-dimensional region (a part of a plane) with a specific intensity in the three-dimensions, Fig. 2.18. The plane has a positive intercept on each axis. On the other hand, WDM and CIM adopt the entire three-dimensional region. Their difference is that WDM separately uses each channel of three-dimensions and CIM exploits the entire three-dimensional region at once. If the three-dimensional channels are orthogonal to each other and can be separated without interference, WDM supported by ISC yields the same performance as that of CIM, otherwise CIM is better than WDM. For analysis of the spectral efficiency, an additive Gaussian noise channel is considered with fully orthogonal optical channels where each receiver can distinguish its corresponding signal, i.e., $S_i = \int r_i(\lambda) I_i(\lambda) d\lambda$. Then, the received signal vector $V = [V_1, \ldots, V_N]^T$ can be denoted by $\mathbf{V} = \mathbf{S} + \mathbf{W}$, where $\mathbf{W}$ is an additive Gaussian noise vector. By the constraints of lighting, the weighted average of $\mathbf{S}$ belongs to a subspace of signal space, which is reduced to a point when $N = 3$, with three color channels.

Figure 2.21 shows one-dimensional mutual information with respect to A/σ and dimming constraints. When the optimal point is chosen in the subspace, the sum of $I(S_i; V_i)$ is the upper-bound for the capacity of CIM. In addition, a two-dimensional example is depicted in Fig. 2.22 for A/σ 8 dB and 6 dB, and dimming targets 0.8 and 0.5 for each dimension, respectively. Two axes represent the signal communicated between corresponding transmitter and receiver. Then, the mutual information is $0.9494 + 0.9385 = 1.8879$ bits/symbol, and the number of symbols is $4 \times 3 = 12$. Thus, the upper-bound for the capacity is 1.8879. A three-dimensional extension is shown in Fig. 2.23 with the lighting constraint of A/σ 5 dB, dimming target 0.3 for the additional dimension. Thus, the mutual information is equal to $0.9494 + 0.9385 + 0.6945 = 2.5824$ bits/symbol.

Table 2.4 compares the mutual information of the CIM and CSK. The channel condition is identical to the case shown in Fig. 2.23 with the exception of the dimming target 0.2 for the first channel R_1 instead of 0.8. The results for CSK1 and CSK2 are obtained to maximize the minimum distance between three symbols without and with consideration of channel gains, respectively. The following three CIM schemes place symbols in a three-dimensional space in different ways. CIM1 places eight equiprobable symbols centered on

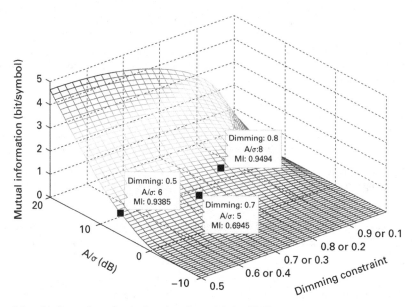

Figure 2.21   Mutual information of a single color channel (after [21]).

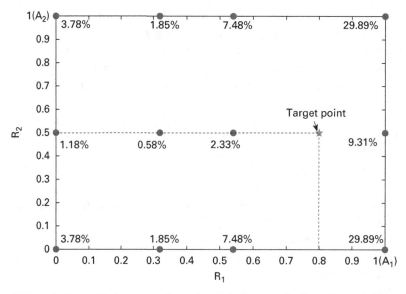

Figure 2.22   CIM symbol constellation in two-dimensional single space for the orthogonal channels: $(A/\sigma,$ dimming) = (8 dB, 80%) in R_1 axis, and $(A/\sigma,$dimming) = (6 dB, 50%) in R_2 axis (after [21]).

the target point so that the set of symbols forms a rectangular parallelepiped. CIM2 locates eight symbols with different probabilities at the corners of the rectangular parallelepiped as in Fig. 2.23. CIM3 denotes the scheme shown in Fig. 2.23. Thus, its signal constellation corresponds to a mirror image of Fig. 2.23 along the R_1 axis.

Table 2.4 Performance comparison of CIM and CSK (© 2012 IEEE, reprinted, with permission, from [21]).

Modulation scheme	Mutual information (bits/symbol)
CSK1-general	1.4768
CSK2-optimal	1.5043
CIM1-analog dimming	2.0070
CIM2-binary	2.3247
CIM3-optimal	2.5824

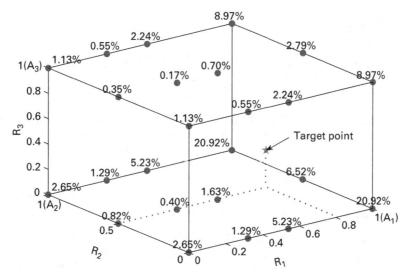

Figure 2.23   CIM symbol constellation in three-dimensional signal space for the orthogonal channels: $(A/\sigma,\text{dimming}) = (8\text{ dB}, 80\%)$ in R_1 axis, $(A/\sigma,\text{dimming}) = (6\text{ dB}, 50\%)$ in R_2 axis, and $(A/\sigma,\text{dimming}) = (5\text{ dB}, 30\%)$ in R_3 axis (after [21]).

Finally, a non-orthogonal multi-colored channel is considered. Since the signals received at the receivers are not independent, the resulting signal subspace forms a parallelogram and a parallelepiped in the two-dimensional and three-dimensional signal space, respectively. Figure 2.24 illustrates the case where the first receiver may respond to the second transmitter. Thus, the receiver signal at the first receiver (R_1') is formed as $R_1' = R_1 + R_2$. On the other hand, the second receiver can distinguish the signal from the second transmitter. Since 2 dB difference occurs between R_1 and R_2, the maximally achieved data rate from R_1' is given as $1 + 10^{-0.2} \approx 1.63$ instead of 2. The corresponding mutual information is 1.9258 bits/symbol and no geometrically meaningful pattern is found among symbols. Figure 2.25 shows the case where the second receiver responds to the first transmitter with only a half responsivity, i.e., $R_2' = 0.5R_1 + R_2$. Then, the corresponding mutual information is equal to 2.0458 bits/symbol.

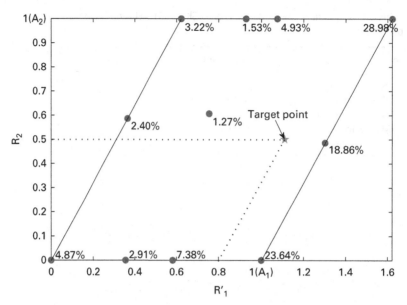

Figure 2.24   Two-dimensional non-orthogonal signal space with $R'_1 = R_1 + R_2$ and R_2: $(A/\sigma, \text{dimming}) = (8\ \text{dB}, 80\%)$ in R_1 axis and $(A/\sigma, \text{dimming}) = (6\ \text{dB}, 50\%)$ in R_2 axis (after [21]).

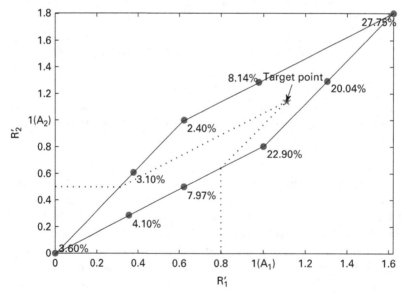

Figure 2.25   Two-dimensional non-orthogonal signal space with $R'_1 = R_1 + R_2$ and $R'_2 = R_2 + 0.5R_1$: $(A/\sigma, \text{dimming}) = (8\ \text{dB}, 80\%)$ in R_1 axis and $(A/\sigma, \text{dimming}) = (6\ \text{dB}, 50\%)$ in R_2 axis (after [21]).

References

[1] S. Boyd and L. Vandenberghe, *Convex Optimization*, Cambridge University Press, 2004.

[2] S. Rajagopal, R. D. Roberts, and S.-K. Lim, "IEEE 802.15.7 visible light communication: Modulation schemes and dimming support," *IEEE Commun. Mag.*, **50**, (*3*), 72–82, 2012.

[3] K. Lee and H. Park, "Modulations for visible light communications with dimming control," *IEEE Photon. Technol. Lett.*, **23**, (*16*), 1136–1138, 15 Aug. 2011.

[4] G. Ntogari, T. Kamalakis, J. W. Walewski, and T. Sphicopoulos, "Combining illumination dimming based on pulse-width modulation with visible-light communications based on discrete multitone," *IEEE/OSA J. Opt. Commun. Netw.*, **3**, (*1*), 56–65, 2011.

[5] W. O. Popoola, E. Poves, and H. Haas, "Error performance of generalised space shift keying for indoor visible light communications," *IEEE Trans. Commun.*, **61**, (*5*), 1968–1976, 2013.

[6] Z. Wang, W.-D. Zhong, C. Yu, *et al.*, "Performance of dimming control scheme in visible light communication system," *Opt. Express*, **20**, (*17*), 18861–18868, 13 Aug. 2012.

[7] E. Cho, J.-H. Choi, C. Park, *et al.*, "NRZ-OOK signaling with LED dimming for visible light communication link," in Proc. 16th European Conference on *Networks and Optical Communications (NOC)*, Newcastle-Upon-Tyne, UK, July 2011, pp. 32–35.

[8] H.-J. Jang, J.-H. Choi, Z. Ghassemlooy, and C. G. Lee, "PWM-based PPM format for dimming control in visible light communication system," in Proc. 8th International Symposium on *Communication Systems, Networks & Digital Signal Processing (CSNDSP)*, Poznan, Poland, July 2012.

[9] S. Arnon, *et al.*, *Advanced Optical Wireless Communication Systems*, Cambridge University Press, 2012.

[10] M. Anand and P. Mishra, "A novel modulation scheme for visible light communication," in Proc. 2010 Annual IEEE India Conference *INDICON*, Kolkata, India, Dec. 2010.

[11] J. K. Kwon, "Inverse source coding for dimming in visible light communications using NRZ-OOK on reliable links," *IEEE Photon. Technol. Lett.*, **22**, (*19*), 1455–1457, 1 Oct. 2010.

[12] K.-I. Ahn and J. K. Kwon, "Capacity analysis of M-PAM inverse source coding in visible light communications," *IEEE/OSA J. Lightw. Technol.*, **30**, (*10*), 1399–1404, 15 May 2012.

[13] A. B. Siddique and M. Tahir, "Joint brightness control and data transmission for visible light communication systems based on white LEDs," in Proc. IEEE *Consumer Communications and Networking Conference*, Las Vegas, NV, Jan. 2011, pp. 1026–1030.

[14] J. Kim, K. Lee, and H. Park, "Power efficient visible light communication systems under dimming constraint," in Proc. 23rd IEEE International Symposium on *Personal Indoor and Mobile Radio Communications (PIMRC)*, Sydney, Australia, Sept. 2012, pp. 1968–1973.

[15] S. H. Lee, K.-I. Ahn, and J. K. Kwon, "Multilevel transmission in dimmable visible light communication systems," *IEEE/OSA J. Lightw. Technol.*, **31**, (*20*), 3267–3276, 15 Oct. 2013.

[16] S. H. Lee and J. K. Kwon, "Turbo code-based error correction scheme for dimmable visible light communication systems," *IEEE Photon. Technol. Lett.*, **24**, (*17*), 1463–1465, 1 Sept. 2012.

[17] S. Kim and S.-Y. Jung, "Novel FEC coding scheme for dimmable visible light communication based on the modified Reed–Muller codes," *IEEE Photon. Technol. Lett.*, **23**, (*20*), 1514–1516, 15 Oct. 2011.

[18] S. Kim and S.-Y. Jung, "Modified RM coding scheme made from the bent function for dimmable visible light communications," *IEEE Photon. Technol. Lett.*, **25**, (*1*), 11–13, 1 Jan. 2013.

[19] P. Das, B.-Y. Kim, Y. Park, and K.-D. Kim, "A new color space based constellation diagram and modulation scheme for color independent VLC," *Advances in Electrical and Computer Engineering*, **12**, (*4*), 11–18, Nov. 2012.

[20] B. Bai, Q. He, Z. Xu, and Y. Fan, "The color shift key modulation with non-uniform signaling for visible light communication," in Proc. 1st IEEE International Conference on *Communications in China* Workshops *(ICCC)*, Beijing, China, Aug. 2012, pp. 37–42.

[21] K.-I. Ahn and J. K. Kwon, "Color intensity modulation for multicolored visible light communications," *IEEE Photon. Technol. Lett.*, **24**, (*24*), 2254–2257, 15 Dec. 2012.

[22] IEEE Standard for Local and Metropolitan Area Networks–Part 15.7: *Short-Range Wireless Optical Communication Using Visible Light*, IEEE Standard 802.15.7–2011, Sept. 2011.

[23] S. Hranilovic and F. R. Kschischang, "Optical intensity-modulated direct detection channels: Signal space and lattice codes," *IEEE Trans. Inf. Theory*, **49**, (*6*), 1385–1399, 2003.

[24] Y. Ohno, "CIE fundamentals for color measurements," Proc. of IS&T NIP16 Intl. Conf. on *Digital Printing Technologies*, Vancouver, Canada, Jan. 2000. pp. 540–545.

3 Performance enhancement techniques for indoor VLC systems

Wen-De Zhong and Zixiong Wang

3.1 Introduction

Light-emitting diodes (LEDs) have been widely deployed for illumination, due to their high performance and energy efficiency properties compared with conventional incandescent and fluorescent lamps [1]. In addition, advantages such as high frequency response, free spectrum license and high security have made LEDs a promising means for wireless communications, whereby people can access the internet through the same visible light. Significant research has been carried out to develop indoor high data rate visible light communication (VLC) systems [1–11]. The data rate of VLC systems in a laboratory environment has been demonstrated to reach the order of Gb/s [12, 13]. In addition, advanced modulation schemes such as spatial modulation [14–17] have been introduced to VLC systems to enhance the data rate considerably. In 2011, light-fidelity (Li-Fi) was introduced by Haas [18], who demonstrated that a VLC system can be leveraged to develop an alternative method of accessing network resources as a substitute for wireless fidelity (Wi-Fi). Although there has been significant progress in this area in the past decade, there are still some challenges to overcome in order to implement and deploy VLC systems [19] on a larger scale. Two major challenges are selection of an uplink transmission approach, and design of energy-saving receivers for long transmission distance. In this chapter, a number of recently proposed techniques [20–25] for enhancing the performance of indoor VLC systems are discussed along with results pertaining to their performance. These include a receiver plane tilting technique [21] and an LED lamp arrangement approach [22, 23] to improve the signal-to-noise ratio (SNR) and bit error rate (BER) performances, and performance evaluation of VLC systems under a dimming control scheme [20, 24].

3.2 Performance improvement of VLC systems by tilting the receiver plane

In a VLC system, the receiver may be located far from the LED lamp, where the SNR is much smaller than at those locations that are close to the lamp. The lower SNR is obtained as the distance from the source increases and as the incident angle increases. This section describes a receiver plane tilting technique [21] to improve the performance of the VLC system across the entire room. The dimension of the room is assumed to be 5 m length × 5 m width × 3 m height. The SNRs with/without tilting the receiver plane are analyzed and compared. For simplicity, the reflections of walls are not considered in analyzing the SNR since the light from the line-of-sight (LOS) is dominant.

3.2.1   SNR analysis of VLC system with a single LED lamp

Figure 3.1 illustrates the geometry of an indoor VLC system with one LED lamp located on the ceiling. The parameters of the VLC system considered in this chapter are given in Table 3.1. The LED lamp is assumed to be located at the center of the ceiling whose position is [2.5 m, 2.5 m, 3.0 m], and the photo-detector (receiver) is on a desk with a height of 0.85 m from the floor. Let φ be the angle of radiation with respect to the axis normal to the LED surface (plane). Following [1, 8], the emitted light from an LED is assumed to have a Lambertian emission pattern, and the radiation pattern is given by:

$$R(\varphi) = \frac{(m+1)\cos^m \varphi}{2\pi}, \qquad (3.1)$$

where m is the order of Lambertian emission, which is related to the transmitter's semi-angle at half power $\varphi_{1/2}$ as $m = \ln(1/2)/\ln(\cos\varphi_{1/2})$.

The frequency response of the LED and photo-detector is assumed to be flat within the modulation bandwidth of the signals considered in this chapter. Considering only the LOS transmission path, the channel DC gain is given by [1, 27]

$$H(0) = R(\varphi)\frac{A}{d^2}\cos\theta = \frac{(m+1)\cos^m \varphi A}{2\pi d^2}\cos\theta, \qquad (3.2)$$

where d is the distance between the LED source and the receiver, A is the physical area of photo-detector, and θ is the angle of incidence with respect to the axis normal to the desk

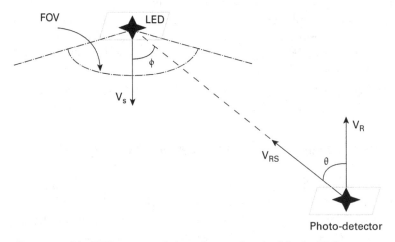

Figure 3.1   Geometry of the LED source and photo-detector (receiver) in the VLC system.

Table 3.1 Parameters of the VLC system considered in this chapter [1, 20–22].

Room size (length × width × height)	5 m × 5 m × 3 m
Height of desk where the receiver is located	0.85 m
Transmitter's semi-angle at half power ($\varphi_{1/2}$)	60 °
Physical area of photo-detector (A)	10^{-4} m^2
Receiver's field of view (FOV)	170 °
Responsivity of photo-detector (R)	1 A/W
Background current (I_{bg})	5100 μA
Noise bandwidth factor (I_2)	0.562
Field-effect transistor (FET) transconductance (g_m)	30 mS
FET channel noise factor (Γ)	1.5
Fixed capacitance (η)	112 pF/cm^2
Open-loop voltage gain (G)	10
Definite integral involved in expression of circuit noise (I_3) [26]	0.0868

plane where the LED is located. Angles φ and θ are associated with the locations/positions of both LED source and receiver. Let $[X_S, Y_S, Z_S]$ and $[X_R, Y_R, Z_R]$ be the locations (coordinates) of source and receiver, respectively. With reference to Fig. 3.1, the radiation angle φ is determined by

$$\cos\varphi = \frac{Z_S - Z_R}{\|[X_S, Y_S, Z_S] - [X_R, Y_R, Z_R]\|}, \tag{3.3}$$

where $\|X\|$ is the norm of X. Equation (3.3) indicates that the radiation angle φ is constant for the given locations of the source and receiver. The value of the incident angle θ is determined not only by the locations of source and receiver, but also by the dihedral angle between the receiver plane and the desk plane where the receiver is located. With reference to Fig. 3.1, let v_{RS} be the vector from the receiver to the source, and v_R be the vector of the receiver. Then the incident angle θ is calculated by

$$\cos\theta = \frac{(v_R, v_{RS})}{\|v_R\| \cdot \|v_{RS}\|}, \tag{3.4}$$

where (v_{RS}, v_R) is the inner product of v_{RS} and v_R. Substituting Eq. (3.4) into Eq. (3.1), the channel DC gain in Eq. (3.1) becomes [21]

$$H(0) = \frac{(m+1)}{2\pi d^2} A \cos^m\varphi \frac{(v_R, v_{RS})}{\|v_R\| \cdot \|v_{RS}\|}. \tag{3.5}$$

Assume that the LED light is modulated with a modulating signal $f(t)$. The optical signal at the output of an LED can be expressed by $p(t) = P_t(1 + M_I f(t))$, where P_t is the launched power of the LED lamp and M_I is the modulation index [28], which is assumed to be 0.2. The received optical power P_r is given by

$$P_r = H(0)P_t. \tag{3.6}$$

After photo-detection and considering that the DC component of the detected signal is filtered out in the receiver, the output electrical signal is given by

$$s(t) = RP_rM_If(t), \tag{3.7}$$

where R is the responsivity of the photo-detector. Hence, the SNR of the output electrical signal can be calculated by [5],

$$SNR = \frac{\overline{s(t)^2}}{P_{\mathrm{noise}}} = \frac{(RH(0)P_tM_I)^2\overline{f(t)^2}}{P_{\mathrm{noise}}}, \tag{3.8}$$

where $\overline{s(t)^2}$ is the average power of the output electrical signal, and P_{noise} is the noise power. The noise power consists of both shot noise and thermal noise, whose variances are given by [1],

$$\sigma_{\mathrm{shot}}^2 = 2q\left[RP_r\left(1 + \overline{\left(M_{\mathrm{index}}f(t)\right)^2}\right) + I_{\mathrm{bg}}I_2\right]B, \tag{3.9}$$

$$\sigma_{\mathrm{thermal}}^2 = 8\pi kT_{\mathrm{K}}\eta AB^2\left(\frac{I_2}{G} + \frac{2\pi\Gamma}{g_{\mathrm{m}}}\eta AI_3B\right), \tag{3.10}$$

where $P_r\left(1 + \overline{\left(M_{\mathrm{index}}f(t)\right)^2}\right)$ is the total received power, q is the electron charge [13], B is the equivalent noise bandwidth, k denotes the Boltzmann constant, and T_{K} represents the absolute temperature. The parameters in Eqs. (3.1)–(3.10) and other parameters used in the VLC system are listed in Table 3.1.

Using Eq. (3.8), the SNR distribution of the receiver located on the desk plane at the height of 0.85 m can be calculated. In calculating the SNR distribution, the parameters in Table 3.1 are used. Figure 3.2 shows the SNR distribution for the case when the launching power of the LED lamp is 5 W, and the LED is located at the center of the ceiling. As shown in Fig. 3.2, the maximum SNR is 28.94 dB when the receiver is located right below the LED, while the minimum SNR is 6.23 dB when the receiver is placed at a corner of the room. Thus, the peak-to-trough SNR difference is 22.70 dB. Note that the SNR vertical tint bar on the right hand side of the figure indicates the relationship between the SNR value and the color (black represents the smallest value of SNR; while white denotes the highest value of SNR). The large variation of the SNR in a room can significantly reduce the overall system performance [21]. The large SNR difference in a room is caused not only by the distance between the LED and the receiver, but also by the non-normal incidence of the light from the LED to the receiver. For the given locations of the LED and receiver, the distance between the LED and the receiver cannot be changed. However, the incident angle of the light from the LED to the receiver can be adjusted to reduce the SNR variation.

3.2.2   Receiver plane tilting technique to reduce SNR variation

As described above, the non-normal incidence of the light could result in a large SNR variation in a room. The incident angle θ is determined by the vectors v_R and v_{RS}. Note that the vector v_R is always perpendicular to the receiver plane. The vector v_{RS} is constant for

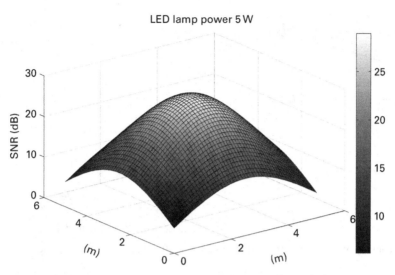

Figure 3.2   SNR distribution of VLC system in a room with one LED lamp located on the center of the ceiling.

the given locations of the source and receiver. According to Eq. (3.4), $\cos\theta$ reaches its maximum when the two vectors $\boldsymbol{v_{RS}}$ and $\boldsymbol{v_R}$ are parallel to each other, i.e., the receiver plane faces the source. In the case where the receiver is not located on the desk right below the source on the ceiling, especially when the receiver is positioned in one of the corners of the room, the maximum channel DC gain is greatly reduced as the incident angle θ increases. However, by tilting the receiver plane towards the source such that the two vectors $\boldsymbol{v_{RS}}$ and $\boldsymbol{v_R}$ become parallel to each other, the value of $\cos\theta$ reaches its maximum and thus maximum channel DC gain of a particular position can be attained, which is only associated with the transmission distance d and the angle of radiation φ [21].

Vector $\boldsymbol{v_{RS}}$ can be expressed as $\boldsymbol{v_{RS}} = [a, b, c] = [X_R, Y_R, Z_R] - [X_S, Y_S, Z_S]$. Here we assume that tilting the receiver plane does not change the position of the receiver. In the spherical coordinate system, the location of the receiver is selected as the origin. Before tilting the receiver plane, the vector $\boldsymbol{V_R}$ is [0, 0, 1], which means that the receiver plane points to the ceiling. After tilting the receiver plane towards the LED source on the ceiling, the vector $\boldsymbol{V_R}$ becomes [$\sin\beta\cdot\cos\alpha$, $\sin\beta\cdot\sin\alpha$, $\cos\beta$], where β is the inclination angle [29] which is equal to the tilting angle, as shown in Fig. 3.3 and the azimuth angle α is determined by the positions of the receiver as well as the source projection on the desk. In the Cartesian coordinate system with the receiver as the origin, the value of angle α is given by Eq. (3.11):

$$\alpha = \begin{cases} \arctan\left(\left|(Y_S - Y_R)/(X_S - X_R)\right|\right) & \text{source projection in the 1st quadrant,} \\ \pi - \arctan\left(\left|(Y_S - Y_R)/(X_S - X_R)\right|\right) & \text{source projection in the 2nd quadrant,} \\ \pi + \arctan\left(\left|(Y_S - Y_R)/(X_S - X_R)\right|\right) & \text{source projection in the 3rd quadrant,} \\ 2\pi - \arctan\left(\left|(Y_S - Y_R)/(X_S - X_R)\right|\right) & \text{source projection in the 4th quadrant.} \end{cases}$$

$$(3.11)$$

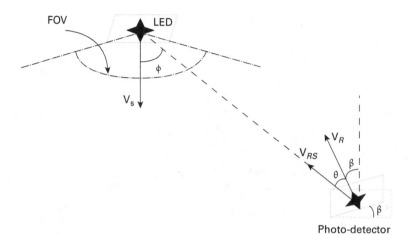

Figure 3.3   Geometry of the LED source and photo-detector after tilting the receiver plane.

Thus, $\cos\theta$ in Eq. (3.4) becomes

$$\cos\theta = \frac{(v_R, v_{RS})}{\|v_R\| \cdot \|v_{RS}\|} = \frac{a\sin\beta\cos\alpha + b\sin\beta\sin\alpha + c\cos\beta}{\sqrt{a^2 + b^2 + c^2}}. \qquad (3.12)$$

Substituting Eq. (3.12) into Eq. (3.5), the channel DC gain after tilting the receiver plane, denoted by $f(\beta)$, becomes,

$$f(\beta) = \frac{(m+1)\cos^m\varphi A}{2\pi d^2\sqrt{a^2 + b^2 + c^2}}(a\sin\beta\cos\alpha + b\sin\beta\sin\alpha + c\cos\beta). \qquad (3.13)$$

Since the receiver is located on the desk, the initial inclination angle β is zero. This receiver plane tilting technique can be implemented by electrical machinery. When the inclination angle is increased after tilting the receiver plane, the two vectors v_R and v_{RS} tend to become parallel to each other and thereby the received optical power increases. The electrical machinery will not stop changing the inclination angle β until the received optical power does not increase any more.

The Newton method (a fast algorithm to find the maximum value of $f(\beta)$ [30]) can be employed to search for the optimum inclination angle β. After finding the optimum tilting angle by the Newton method, the maximum optical power is obtained for each receiver position. Figure 3.4 shows the improved SNR distribution. The maximum SNR remains unchanged at 28.94 dB, while the minimum SNR at each corner of the room is increased to 11.92 dB, resulting in an improvement of 5.69 dB in the peak-to-trough SNR difference as compared to the case without tilting the receiver plane.

3.2.3   Multiple LED lamps with the receiver plane tilting technique

To further reduce the SNR variation, multiple LED lamps could be employed in conjunction with the receiver plane tilting technique [21]. As an example, consider a case where four LED lamps are located on the ceiling at the positions of [1.5 m, 1.5 m,

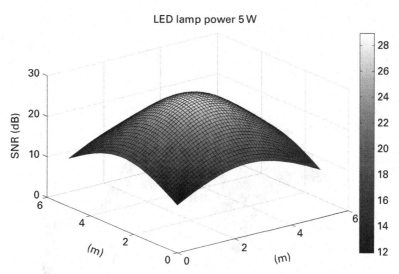

LED lamp power 5 W

Figure 3.4 SNR distribution of VLC system in a room with one LED lamp on the center of the ceiling after tilting the receiver plane.

3.0 m], [1.5 m, 3.5 m, 3.0 m], [3.5 m, 1.5 m, 3.0 m] and [3.5 m, 3.5 m, 3.0 m], respectively. As the light is received from all of the four LED lamps, the channel DC gain in Eq. (3.2) is modified as follows:

$$H(0) = \sum_{i=1}^{4} \frac{(m+1)A\cos^m \varphi_i}{2\pi d_i^2} \cos\theta_i, \tag{3.14}$$

where the subscript i denotes lamp i. In this case, the total radiating power from the four LEDs remains unchanged at 5 W, i.e., the launching power of each LED lamp is reduced to one quarter of that of one LED lamp in subsection 3.2.2. Figure 3.5 (a) shows the SNR distribution with four LED lamps without tilting the receiver plane. As shown in Fig. 3.5 (a), the maximum and minimum SNRs are 22.72 dB and 8.95 dB, respectively. That is, the peak-to-trough SNR difference is 13.77 dB without tilting the receiver plane. It is also observed that in the area inside the projections of the LEDs on the desk plane, the SNR distribution is almost constant and hence there is no need to adjust the SNR distribution within this area [21].

However, the SNR variation is quite large in the places outside projections of the four LEDs on the desk plane. In such places, the SNR difference can be reduced by tilting the receiver plane in the same way as described for the case of a single LED lamp.

When the receiver is not equidistant with respect to any of two LEDs, it faces the nearest LED of the four, which determines the value of azimuth angle α. When the receiver is equidistant from two LEDs, it faces to the middle of them. The total channel DC gain after tilting the receiver plane, denoted as $f(\beta)$, is given by

$$f(\beta) = \frac{(m+1)A\cos^m \varphi}{2\pi d^2 \sqrt{a^2 + b^2 + c^2}} (a\sin\beta\cos\alpha + b\sin\beta\sin\alpha + c\cos\beta). \tag{3.15}$$

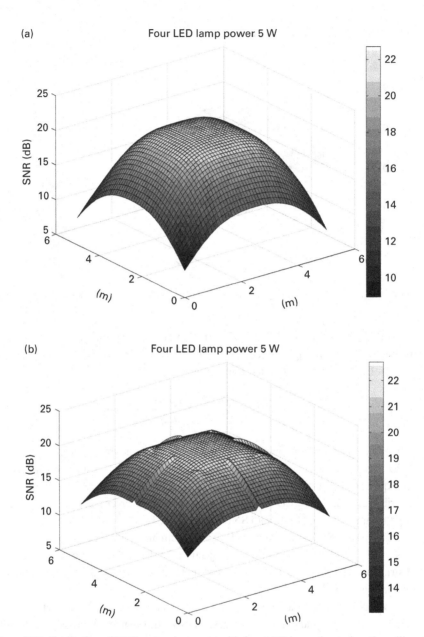

Figure 3.5   SNR distribution of VLC system in a room with four LED lamps located on the ceiling: (a) before; and (b) after tilting the receiver plane (after [21]).

As in the case of a single LED lamp, the optimum tilting angle β can then be obtained by the Newton method. Figure 3.5 (b) shows the improved SNR distribution with the receiver tilting technique. As shown in Fig. 3.5 (b), the maximum SNR remains at 22.72 dB, while the minimum SNR increases to 13.09 dB. That is, the peak-to-trough SNR difference is reduced from 13.77 dB to 9.63 dB. In other words, a 4.14 dB improvement in the peak-to-trough SNR difference is achieved. In the case of four LED lamps, the study shows that only three search steps are required by the Newton algorithm to converge to the optimum value [21].

3.2.4   Spectral efficiency

Tilting the receiver plane makes it possible to attain optimum SNR. A higher SNR per symbol (E_s/N_0) means a better bit error rate (BER) performance. As discussed above, the SNR can vary considerably within the room. Similarly to RF wireless communications, adaptive advanced modulation formats such as M-ary quadrature amplitude modulation (M-QAM) orthogonal frequency division multiplexing (OFDM) can be employed to enhance transmission capacity [31]. This subsection discusses the spectral efficiency of a single user VLC system employing adaptive M-QAM OFDM. Figure 3.6 is a block diagram for a VLC system with adaptive M-QAM OFDM, where the value of M represents the number of points in the signal constellation and can be varied in accordance with the SNR. Here it is assumed that infrared (IR) or another kind of wireless technology is employed to provide channel feedback as well as uplink transmission.

Following [32, 33] and applying gray coding for the mapping of the M-QAM signal, the BER of the M-QAM OFDM signal is given by

$$\mathrm{BER} \approx \frac{4}{\log_2(M)}\left(1 - \frac{1}{\sqrt{M}}\right)Q\left(\sqrt{\frac{3\log_2(M)}{M-1}\frac{E_b}{N_0}}\right), \qquad (3.16)$$

where $Q(\cdot)$ is the Q-function. Note that the relationship between SNR per symbol (E_s/N_0) and SNR per bit (E_b/N_0) is given by $E_s/N_0 = \log_2(M) \times E_b/N_0$ [33].

As shown in Fig. 3.6, when the M-QAM OFDM optical signal reaches the photodetector, its power is detected and sent back to the sources on the ceiling via the IR feedback channel after tilting the receiver plane. Here the benchmark BER is set at 10^{-3}, which satisfies the requirement of error-free transmission by applying a forward-error correction (FEC) code [34]. A smaller value of M should be chosen to attain a BER of 10^{-3} at the locations/positions with low SNR. However, at the places with high SNR, a larger value of M is selected to achieve a higher data rate while maintaining a steady BER of 10^{-3}. It is noted that the value of the M-QAM OFDM signal should be real in optical transmission (this is attained by applying Hermitian symmetry [4]), resulting in a 50% reduction in spectral efficiency [35, 36]. Using Eq. (3.16), the SNR per symbol threshold for achieving a BER of 10^{-3} can be calculated for a given value of M, as shown in Table 3.2, where the symbol rate is 50 Msymbol/s.

Let N be the number of subcarriers used in OFDM. Assuming that the pulse shape is rectangular, the spectral efficiency (SE) in units of bit/s/Hz of an M-QAM OFDM signal can be expressed as [37]

$$SE = \frac{1}{2}\log_2(M)\frac{N}{N+1} \approx \frac{1}{2}\log_2(M), \qquad (3.17)$$

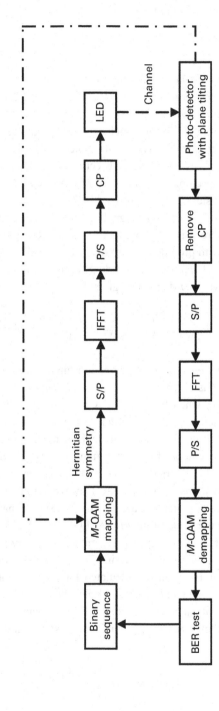

Figure 3.6 Block diagram of a VLC system employing adaptive *M*-QAM OFDM. CP: cyclic prefix, P/S: parallel to serial, S/P: serial to parallel (adapted from [21]).

Table 3.2 The calculated SNR per symbol threshold for achieving a BER of 10^{-3} for different values of M in M-QAM.

Value of M in M-QAM	4	16	64	256	1024
SNR (dB) per symbol threshold	9.8	16.5	22.5	28.4	34.2

where ½ represents the SE reduction as a result of applying the Hermitian symmetry. In a single-user VLC system using adaptive M-QAM OFDM, the value of M is varied in accordance with the value of SNR. The average SE across the entire room can be expressed by

$$\overline{SE} = \frac{1}{2} \sum_i \log_2(M_i)\, p(M_i), \tag{3.18}$$

where $p(M_i)$ is the probability that M_i-QAM is used, which can be calculated based on the SNR distribution in the room.

Figure 3.7 (a) and (b) show the average SE for the cases of one LED lamp and four LED lamps, respectively. The average SE increases with the total LED power. This is because SNR is increased with the total LED power, which in turn increases the probability of employing a larger value of M in adaptive M-QAM modulation. Figure 3.7 (a) and (b) also show the average SE improvement attained by tilting the receiver plane. In the case of one LED lamp, the average improvement is about 0.36 bit/s/Hz, and the maximum improvement is 0.47 bit/s/Hz when the LED lamp power is 9 W. In the case of four LED lamps, the average improvement is 0.18 bit/s/Hz, and the maximum improvement is 0.23 bit/s/Hz when the total LED lamp power is 19 W.

3.3   Performance improvement of VLC systems by arranging LED lamps

For an indoor VLC system, equal signal quality in terms of SNR and BER across the entire room is important; this is particularly true when there are multiple users in the room. As discussed in Section 3.2, the LED lamps are usually located around the center of the ceiling (called a centered-LED lamp arrangement) in a typical room. This centered-LED lamp arrangement makes SNR vary largely from one location/position to another [1, 21], which can significantly affect the quality of the received signal across the room. Section 3.2 analyzes and discusses the performance improvement in an indoor VLC system by tilting the receiver plane. Tilting the receiver plane can reduce the SNR variation to some extent, but it may considerably increase the complexity of the receiver design. This section describes an effective LED lamp arrangement reported in [22] to significantly reduce SNR variation and hence improve BER performance of VLC systems throughout the entire room, so that multiple users can receive signals of almost equal quality, regardless of their location.

(a)

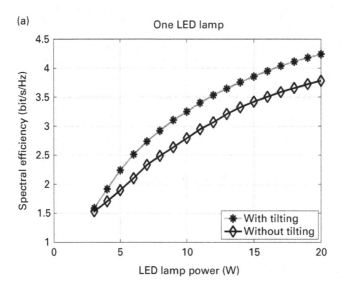

(b)

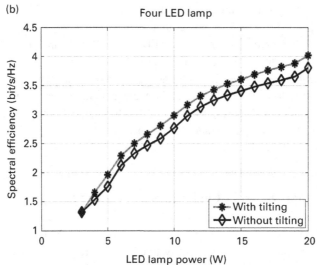

Figure 3.7   Average spectral efficiency for the cases of: (a) one LED lamp; and (b) four LED lamps, with and without tilting receiver plane (adapted from [21]).

3.3.1   Arrangement of LED lamps

In order to see the effectiveness of the LED lamp arrangement, without loss of generality, let us first consider a situation where there are 16 identical LED lamps that are located around the center of the ceiling. The interval between adjacent LED lamps is 0.2 m. Each LED lamp emits 125 mW, and hence the total power of 16 centered-LED lamps is 2 W. The other parameters of the VLC system considered here are provided in Table 3.1. 100 positions are sampled in the room. These sampled positions are uniformly distributed on the plane where the photo-detector is located. To evaluate the quality of signal received at every location in the room, a parameter denoted Q_{SNR} is introduced, which is defined as

$$Q_{SNR} = \frac{\overline{SNR}}{2\sqrt{\mathrm{var}(SNR)}}, \tag{3.19}$$

where $\overline{SNR}$ is the mean of SNR, and $\mathrm{var}(SNR)$ denotes the variance of SNR. A higher value of Q_{SNR} implies that the SNR is more uniformly distributed over the entire room. Figure 3.8 (a) illustrates the calculated SNR distribution, in which the maximum variation in SNR is about 14.5 dB. In this case, the associated Q_{SNR} is approximately 0.5 dB, which means that the SNR varies significantly throughout the room and the signal quality

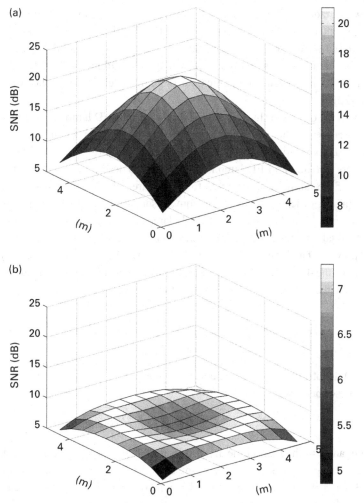

Figure 3.8   SNR distribution of a VLC system in a room with a total power of 2 W: (a) 16 centered-LED lamps; (b) 16 circle-LED lamps (adapted from [22]).

is strongly related to the user's location/position. In calculating SNR distribution, the reflections of walls or the ISI are not considered here, since the LOS light is dominant. However, they will be included in the analysis of BER performance in the next subsection.

The poor SNR distribution illustrated in Fig. 3.8 (a) arises from the centered-LED lamp arrangement, where the distances between a user in the corner of the room and the lamps are much greater than the distances between a user in the center of the room and the lamps.

If the LED lamps are located separately and symmetrically to the center of the ceiling, the SNR distribution is expected to be improved considerably, since the differences in distances between users and lamps are reduced [22]. In [22], a circle-LED lamp arrangement is proposed to reduce SNR variation. Figure 3.8 (b) shows the SNR distribution for the case of 16 lamps, where the LED lamps are distributed evenly on a circle on the ceiling with a radius of 2.5 m and each lamp emits the same amount of power as that in the case of 16 centered-LED lamps. As shown in Fig. 3.8 (b), the SNR variation is reduced largely from 14.5 dB to 2.4 dB, and the Q_{SNR} increases substantially from 0.5 dB to 9.3 dB. This demonstrates that the circle arrangement offers much better signal quality and thereby an improved communication system, which is little related to the user's position [22].

Although the Q_{SNR} is as high as 9.3 dB for the circle-LED lamp arrangement, the SNR in the four corners (which are equivalent in terms of distance) is still smaller than that in other locations, as shown in Fig. 3.8 (b). A technique to further improve the SNRs at the four corners is to add an LED lamp in each corner as described in [22]. Assume that the distances of the lamps in the corners to their nearest walls are all 0.1 m, as shown in Fig. 3.9 (a). In order to make a fair comparison, the total LED lamp power is still maintained at 2 W and the number of LED lamps placed in the circle is reduced to 12, so that the total number of LED lamps remains unchanged. Let $P_{t,circle}$ and $P_{t,corner}$ be the emitted powers of a circle-LED lamp and a corner-LED lamp, respectively. By adjusting the power of the 4 corner-LED lamps and the 12 circle-LED lamps spreading uniformly on the circle, the variance of the received optical power P_r, can be minimized, as follows [22]:

$$\min \text{var}(P_r) = \min E[(P_{r,j} - E(P_{r,j}))^2], \tag{3.20}$$

where $E(\cdot)$ denotes the mean value and $P_{r,j}$ is the received power at sampled position j, which is described by

$$P_{r,j} = \sum P_{t,corner} H(0)_{corner} + \sum P_{t,circle} H(0)_{circle}. \tag{3.21}$$

Next, the radius of the LED circle and the distance between a corner-LED and its nearest walls are changed to find the minimum SNR fluctuation. The results are given in Tables 3.3 and 3.4. As shown in Table 3.3, when the distance between the corner-LED lamps and their nearest walls is 0.1 m, the optimum radius of the LED circle is between 2.2 to 2.3 m, where Q_{SNR} is largest. Table 3.4 shows that the optimum distance between

Table 3.3 SNR and Q_{SNR} under the arrangement of 12 circle-LEDs and 4 corner-LEDs with different radii, where the distance between a corner-LED and its nearest walls is 0.1 m [22].

Radius (m)	2.1	2.2	2.3	2.5
SNR range (dB) [min, max]	[5.5, 6.4]	[5.6, 6.5]	[5.5, 6.5]	[5.3, 6.5]
Q_{SNR} (dB)	12.1	12.2	12.2	11.5

Table 3.4 SNR and Q_{SNR} under the arrangement of 12 circle-LEDs and 4 corner-LEDs with different distances between a corner-LED and its nearest walls, where the radius of the LED circle is 2.2 m [22].

Distance (m)	0.5	0.25	0.15	0.1
SNR range (dB) [min, max]	[6.0, 7.4]	[5.9, 6.8]	[5.7, 6.6]	[5.6, 6.5]
Q_{SNR} (dB)	10.7	11.7	12.1	12.2

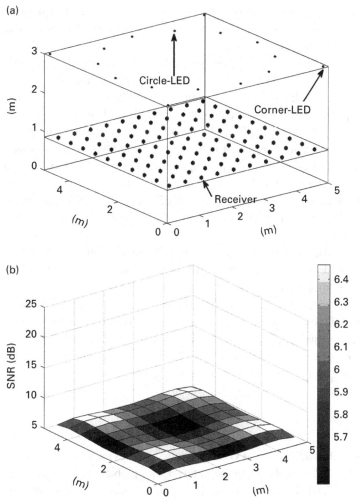

Figure 3.9   Arrangement of 12 LED lamps in a circle and 4 LED lamps in corners: (a) locations of LED lamps and 100 receivers; (b) SNR distribution with a total power of 2 W (adapted from [22]).

the corner-LED lamps and their nearest walls is 0.1 m, when the radius of the LED circle is 2.2 m. It is also found that the largest Q_{SNR} is obtained when the powers of each corner-LED lamp and each circle-LED lamp are 238 mW and 87 mW, respectively. The improved SNR distribution is shown in Fig. 3.9 (b), where the maximum SNR difference is 0.85 dB and the corresponding Q_{SNR} is 12.2 dB. The above results reveal that the arrangement of 12 circle-LED lamps and 4 corner-LED lamps with the given parameters in Table 3.1 can provide almost identical communication quality to multi-users, irrespective of their positions in the room [22].

3.3.2   BER analysis

As discussed above, the arrangement of 12 circle-LED lamps and 4 corner-LED lamps can provide almost uniformly distributed SNR to users, irrespective of their locations/positions in the room. However, this may cause an increase in ISI since the photodetector receives signals from all of the LED lamps whose distances to the receiver vary greatly, which may reduce the BER performance significantly. Without considering reflections, the maximum difference of light arrival time under the arrangement of 12 circle-LED lamps and 4 corner-LED lamps is 15.9 ns when the receiver is located in a corner position [22]; whereas the maximum time difference is only 2.34 ns when the 16 LED lamps are located in the center of the ceiling. In [22], the BER performance of both 100 Mb/s and 200 Mb/s bipolar OOK signals is analyzed and evaluated.

This subsection presents the BER analysis of a 100 Mbit/s bipolar OOK signal under the arrangement of 12 circle-LED lamps and 4 corner-LED lamps. The first order of reflection is considered in the BER analysis. Let us consider the worst case where the receiver is placed in a corner with a location of [0.25 m, 0.25 m, 0.85 m] and hence the ISI is most severe. The reflectivity and the modulation index are assumed to be 0.7 and 0.2, respectively.

Figure 3.10 (a) illustrates the pulse shape of the received bit "1" under the arrangement of 12 circle-LED lamps and 4 corner-LED lamps, as shown in Fig. 3.9 (a). As can be seen in Fig. 3.10 (a), the duration of the received bit "1" is more than 30 ns, more than three times the transmitted bit period $T = 10$ ns. Let $h = [1\, a_1 \ldots a_k]$ be the normalized channel response. It is noted that $a_i\, (i = 1, 2, \ldots, k)$ represents the ISI contribution from the present bit to the subsequent ith bit. Let I_m be the present received bit with amplitude $\pm\sqrt{E_b}$, and $I_{m-1}, \ldots, I_{m-k}$ be the k preceding received bits. Considering the ISI from the k preceding bits, the present received signal y_m is expressed by

$$y_m = I_m + \sum_{i=1}^{k} a_i\, I_{m-i} + n, \qquad (3.22)$$

where n is the additive white Gaussian noise (AWGN) with power spectral density of $N_0/2$. Let $P(e\,|\,I_m = \sqrt{E_b})$ be the conditional error probability when the present received bit is "1," i.e., amplitude is $\sqrt{E_b}$. Then $P(e\,|\,I_m = \sqrt{E_b})$ can be expressed by [32],

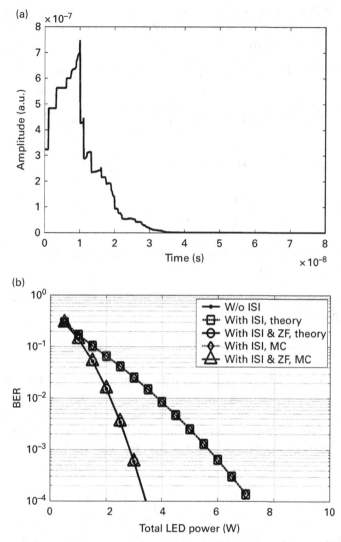

Figure 3.10   BER performance of a 100 Mbit/s bipolar OOK signal when the receiver is located at the corner: (a) the pulse shape of received bit "1" with ISI; and (b) BER with and without ZF equalization, when the total LED power is 2 W (adapted from [22]).

$$P(e \mid I_m = \sqrt{E_b}) = \sum P(I_{m-1}, \ldots, I_{m-k})P(e \mid I_m = \sqrt{E_b}, I_{m-1}, \ldots, I_{m-k}), \quad (3.23)$$

where $I_{m-1}, \ldots, I_{m-k}$ is a combination of the k preceding received bits, and $I_{m-i} (i = 1, 2, \ldots, k) = \pm\sqrt{E_b}$, i.e., each of the k preceding received bits is either 1 or 0. Note that bit 0 is mapped to -1 for a bipolar OOK signal. $P(I_{m-1}, \ldots, I_{m-k})$ is the probability of one such combination. $P(e \mid I_m = \sqrt{E_b}, I_{m-1}, \ldots, I_{m-k})$ is the conditional error probability when the present received bit is 1 and one of the combinations of the k preceding received bits occurs. For example, when all the k preceding received bits and the present received bit are 1, the conditional error probability in Eq. (3.23) is given by

$$
\begin{aligned}
P(e \,|\, I_m &= \sqrt{E_b},\ I_{m-1} = I_{m-2} = \,\ldots\, = I_{m-k} = \sqrt{E_b}) \\
&= P(y_m < 0 \,|\, I_m = I_{m-1} = I_{m-2} = \,\ldots\, = \sqrt{E_b}) \\
&= P\!\left(y_m = \sqrt{E_b}\left(1 + \sum_{i=1}^{k} a_i\right) + n < 0\right) \\
&= P\!\left(n < -(1 + \sum_{i=1}^{k} a_i)\sqrt{E_b}\right) \\
&= Q\!\left((1 + \sum_{i=1}^{k} a_i)\sqrt{2E_b/N_0}\right),
\end{aligned}
\tag{3.24}
$$

where $Q(\cdot)$ is the Q-function. Since the occurrence of bit 1 and bit 0 is equal, the overall BER performance is given by

$$
\begin{aligned}
P(e) &= P(I_m = \sqrt{E_b})P(e \,|\, I_m = \sqrt{E_b}) + P(I_m = -\sqrt{E_b})P(e \,|\, I_m = -\sqrt{E_b}) \\
&= P(e \,|\, I_m = \sqrt{E_b}) \\
&= \sum P(I_{m-1}, \,\ldots\, , I_{m-k})P(e \,|\, I_m = \sqrt{E_b},\ I_{m-1}, \,\ldots\, , I_{m-k}).
\end{aligned}
\tag{3.25}
$$

Due to the ISI from the k preceding bits, BER performance is expected to be reduced significantly without performing equalization on the received signal. Several signal equalization techniques have been developed to mitigate ISI [32]. Here the time domain zero-forcing (ZF) equalization is used to suppress ISI [32, 38]. Let $\{c_n\}$ be the coefficient of the ZF equalizer, and $\{q_n\}$ be the output of the equalizer. Then $\{q_n\}$ is the convolution of $\{c_n\}$ and the channel response h. Ideally, $\{q_n\}$ should be given by

$$
q_n = \sum_{m=-\infty}^{\infty} c_m h_{n-m} = \begin{cases} 1 & (n = 0) \\ 0 & (n \neq 0) \end{cases}.
\tag{3.26}
$$

For a non-ideal equalizer with a finite number of taps, $q_n \neq 0$ when $n \neq 0$, i.e., residual ISI exists. Replacing $h = [1\ a_1 \ldots a_k]$ in Eq. (3.22) with $\{q_n\}$ and substituting y_m into Eq. (3.25), the improved BER performance with time domain ZF equalization is attained.

Figure 3.10 (b) shows the BER performance of both the theoretical analysis and Monte-Carlo (MC) simulation with and without ZF equalization. The theoretical results agree with the simulation results very well (they are completely overlapped in the figure); the BER performance is significantly improved after applying ZF equalization. As an example, the required total LED power is 6.2 W for achieving a BER of 5×10^{-4} without ZF equalization, while the required total LED power is reduced to 3.0 W with ZF equalization. Thus ZF equalization provides 3.2 dB improvement in power reduction. It is also observed that the BER performance with ZF equalization is almost the same as that without ISI (the two curves are completely overlapped). This means that the ISI can be totally mitigated by ZF equalization. By applying an FEC code, error-free transmission could be attained under this BER requirement [34].

3.3.3   Channel capacity analysis

For a noisy communication channel, the channel capacity is defined as the maximum mutual information between the input and output of the channel over the input distribution [38]. For a discrete-input and continuous-output VLC channel, the channel capacity can be expressed by

$$
\begin{aligned}
C &= \max_{P_X(\cdot)} I\,(X;Y) \\
&= \max_{P_X(\cdot)} \sum_{x \in X} P_X(x) \int_{-\infty}^{\infty} f_{Y|X}(y\,|\,x) \log_2 \frac{f_{Y|X}(y\,|\,x)}{f_Y(y)}\,dy,
\end{aligned}
\tag{3.27}
$$

where $P_X(.)$ is the input distribution, x is the discrete-input symbol in the set of X, $f_Y(y)$ is the probability density function of the continuous-output signal y, and $f_{Y|X}(y|x)$ is the conditional probability density function of y when the given input symbol is x. Assuming that the input-discrete signal is a bipolar OOK signal, $f_{Y|X}(y|x)$ is given by,

$$
\begin{cases}
f_{Y|X}(y\,|\,x=-1) = \dfrac{1}{\sqrt{2\pi}\sigma_N} \exp\left(-\dfrac{(y+1)^2}{2\sigma_N^2}\right), \\[3mm]
f_{Y|X}(y\,|\,x=+1) = \dfrac{1}{\sqrt{2\pi}\sigma_N} \exp\left(-\dfrac{(y-1)^2}{2\sigma_N^2}\right),
\end{cases}
\tag{3.28}
$$

where σ_N^2 is the variance of noise. Considering the ISI discussed in Section 3.3.2, the conditional probability density function in Eq. (3.28) can be expressed by

$$
\begin{aligned}
& f_{Y|X}(y_m\,|\,x_m) \\
&= \sum p(x_{m-1},\ \ldots,x_{m-k}) f_{Y|X}(y_m\,|\,x_m,\ x_{m-1},\ \ldots,x_{m-k}) \\
&= \sum \frac{p(x_{m-1},\ \ldots,x_{m-k})}{\sqrt{2\pi}\sigma_N} \exp\left(-\frac{\left(y-[x_m,\ x_{m-1},\ \ldots,x_{m-k}][1,\ a_1,\ \ldots,a_k]'\right)^2}{2\sigma_N^2}\right),
\end{aligned}
\tag{3.29}
$$

where $h = [1\ a_1 \ldots a_k]$ is the channel response, x_m is the present transmitted bit and $x_{m-1},\ldots,x_{m-k}$ are the k previous transmitted bits. Substituting Eq. (3.29) into Eq. (3.27), the channel capacity of a bipolar OOK signal with ISI in a VLC system can be calculated.

Figure 3.11 shows the calculated channel capacity of a 100 Mbit/s bipolar OOK signal versus total LED lamp power under the arrangement of 12 circle-LED lamps and 4 corner-LED lamps with and without ZF equalization. As shown in Fig. 3.11, the ZF equalization improves the channel capacity significantly. The maximum channel capacity improvement by ZF equalization is 0.17 bits/symbol, which occurs at the total LED power of 2 W. It is also shown in the figure that with ZF equalization, the channel

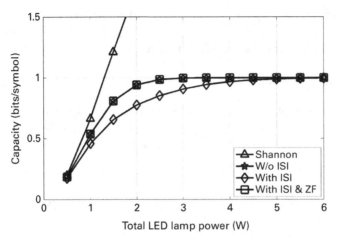

Figure 3.11   Channel capacity of a 100 Mbit/s bipolar OOK signal under the arrangement of 12 circle-LED lamps and 4 corner-LED lamps (adapted from [22]).

capacity remains almost the same as that of the channel without ISI. As a reference, the Shannon capacity is also depicted in Fig. 3.11.

3.4   Dimming control technique and its performance in VLC systems

Illumination and communication are the two major functions of the LED lamps in VLC systems. The brightness of the LED light needs to be adjusted in accordance with the requirements and comfort of users. In addition, adjusting the brightness of LED lamps helps save energy [39]. Pulse width modulation (PWM) is widely used as a dimming control technique [39, 40], where the brightness of the LED light is changed by adjusting the duty cycle of the PWM signal without varying the LED current [20].

As shown in Fig. 3.12 (a), the LED current is modulated by a PWM signal to control its brightness by changing the *on* duration within the whole period. So the light is dimming during the whole period of the PWM signal. The data are modulated onto the dimming-controlled light in the *on* time only, and no light is transmitted during the *off* time as depicted in Fig. 3.12 (b). Since the LED current remains constant throughout, the brightness of the LED light is varied by applying the PWM dimming control signal to adjust the duration of its *on* period as a fraction of the whole period.

With the duty cycle of the PWM signal set to 1, all the LED light is transmitted and the light obtained is of the highest brightness. When the duty cycle is reduced, the LED light in the *off* period is blocked and hence the light is dimming during the whole duration of the PWM signal. It is worth mentioning that the frequency of the PWM signal should be reasonably high, say higher than 200 Hz [41], otherwise it will cause flicker and would have adverse effects on user health [42].

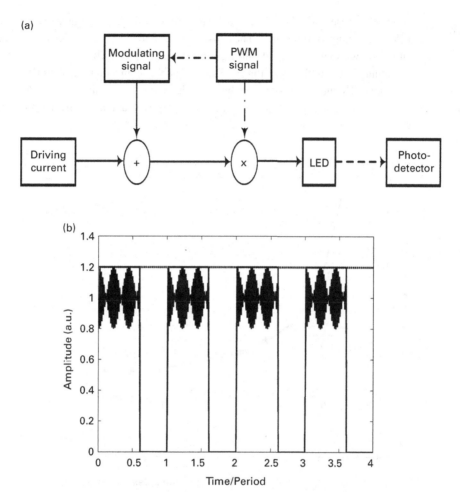

Figure 3.12   (a) Dimming control; (b) signal waveform with dimming control, duty cycle = 0.6   (adapted from [20]).

3.4.1   Bipolar OOK signal under dimming control

As discussed above, with the dimming control, the duration of data transmission in one PWM dimming control signal period (T) is reduced, compared with the case without dimming control. Although the BER performance in the *on* period of the PWM dimming control signal is not changed for a given modulation format, the number of transmitted bits would be reduced in the whole PWM period, which in fact reduces the average data rate. To deal with this problem, the data rate should be increased accordingly while the LED light is dimming, so that the number of transmitted bits remains unchanged. That is, the following equation should hold:

$$R_1 TD = R_0 T, \tag{3.30}$$

where R represents the bit rate of a bipolar OOK signal, and D denotes the duty cycle of a PWM dimming control signal. The subscripts "0" and "1" in Eq. (3.30) correspond to the situations without and with dimming control, respectively. The adaptive data rate under dimming control is shown in Fig. 3.13 (a). In the following analysis, the original data rate R_0 of an OOK signal is assumed to be 10 Mbit/s without dimming control. As depicted in Fig. 3.13 (a), under the dimming control, the adaptive data rate is inversely-proportional

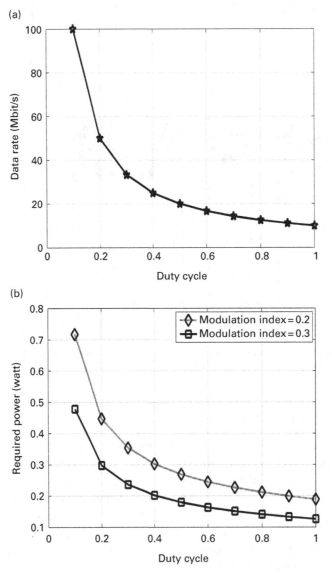

Figure 3.13   (a) Adaptive data rate versus the duty cycle; (b) the required LED lamp power to achieve BER of 10^{-3} without applying dimming control in an OOK VLC system versus the duty cycle (adapted from [20]).

to the duty cycle of the PWM dimming control signal in order to make the number of transmitted bits unchanged. Hence, the adaptive data rate R_1 is higher than the original data rate R_0 and is increased as the duty cycle is reduced. For example, when the duty cycle is 0.1, the adaptive data rate would be 10 times as high as the original data rate, which would make the system hard to implement on the original circuit [20].

Following [43], the BER performance of a bipolar OOK signal is given by

$$BER = Q(\sqrt{2SNR}), \tag{3.31}$$

where $Q(\cdot)$ is the Q-function. As described above, the signal power is kept constant while changing duty cycle; however, the noise power is increased as the data rate grows [1] when the duty cycle decreases, which reduces the BER performance in terms of SNR as shown in Eq. (3.31). Under the dimming control, the BER of less than 10^{-3} should be guaranteed to achieve error-free transmission by applying the FEC code [34]. Substituting Eq. (3.7) into Eq. (3.31), we have

$$BER = Q(\sqrt{2SNR}) = Q\left(\frac{\sqrt{2}RH(0)P_tM_I}{\sigma(P_t)}\right). \tag{3.32}$$

Note that the average power $\overline{f(t)^2}$ of the bipolar OOK signal in Eq. (3.7) is unity and the noise variance σ^2 is related to the total received optical power P_r in terms of the LED lamp power P_t. By solving Eq. (3.32), the required LED lamp power to achieve a BER of 10^{-3} without applying dimming control can be obtained, which is associated with both modulation index M_I and the noise variance, when the locations of LED lamp and receiver are fixed.

In the following scenario, the locations of an LED lamp and receiver are assumed to be [2.5 m, 2.5 m, 3.0 m] and [3.75 m, 1.25 m, 0.85 m], respectively. In addition, the illuminance is assumed to be proportional to the LED's driving current. It is noted that the required LED lamp power to achieve a BER of 10^{-3} should be kept unchanged under the dimming control as well [20]. Figure 3.13 (b) shows the required LED lamp power to achieve a BER of 10^{-3} without applying dimming control versus the duty cycle. When the duty cycle is varied from 1 to 0.3, i.e., the illuminance of LED light is reduced to 30% of the initial illuminance, the required LED lamp powers without applying dimming control increase slowly to 0.35 W and 0.24 W for modulation indices of 0.2 and 0.3, respectively. However, as the duty cycle is further reduced from 0.3 to 0.1, i.e., the illuminance of the LED light is only 10% of the initial illuminance, the required LED lamp power without applying dimming control grows drastically from 0.35 W to 0.72 W when the modulation index is 0.2 and grows from 0.24 W to 0.48 W when the modulation index is 0.3. Since the LED lamp power has to be kept constant throughout the dimming control scheme, in order to achieve a BER of 10^{-3} for the entire duty cycle range of 0.1 to 1, the required LED lamp power should be set as high as 0.72 W if the modulation index is 0.2, and 0.48 W if the modulation index is 0.3. In this way, while the dimming control is applied, the LED lamp power is maintained constant and both average data rate and the BER of 10^{-3} can be guaranteed. The above results reveal that in an OOK VLC system with dimming control, not only the data rate but also the lamp power need to be increased significantly in order to provide a guarantee of communication quality in terms of BER and average data rate for the entire duty cycle range of 0.1 to 1.

3.4.2   Adaptive M-QAM OFDM signal under dimming control

This subsection analyzes and discusses the performance of an adaptive M-QAM OFDM signal under the dimming control scheme, where M represents the number of points in the signal constellation [44]. Since one symbol of the M-QAM signal carries $\log_2(M)$ bits, the total number of transmitted bits could be kept the same by increasing the symbol rate and/or the value of M when higher level M-QAM is used. Let M_0 be the initial number of points in the signal constellation, and M_1 be the adaptive number of points in the signal constellation. Hence, we have

$$\log_2(M_1)R_1 TD = \log_2(M_0)R_0 T,$$
$$R_1 = \frac{\log_2(M_0)R_0}{\log_2(M_1)D}, \tag{3.33}$$

where R_0 is the original symbol rate of the M-QAM signal without dimming control, and R_1 is the adaptive symbol rate of the M-QAM signal with dimming control. Here R_0 is assumed to be 10 Msymbol/s. Substituting SNR per symbol in Eq. (3.8) into Eq. (3.16), the BER performance of an M-QAM signal without applying dimming control is obtained from,

$$BER \approx \frac{4}{\log_2(M)}\left(1 - \frac{1}{\sqrt{M}}\right)Q\left(\sqrt{\frac{3}{M-1}\frac{\left(RH(0)P_t M_I\right)^2}{\sigma^2(P_t)}}\right). \tag{3.34}$$

The average power $\overline{f(t)^2}$ of the M-QAM signal in Eq. (3.8) is also unity. Solving Eq. (3.34), the required LED lamp power to achieve a BER of 10^{-3} without applying dimming control of the M-QAM signal is obtained. Note that when OFDM is applied in a VLC system, the bandwidth should be at least twice as large as the symbol rate, due to the application of Hermitian symmetry. In addition, Eqs. (3.16) and (3.34) are the BER performance for an M-QAM with a square constellation. For an M-QAM with non-square constellation, the threshold of SNR per symbol to achieve a BER of 10^{-3} is obtained by using MC simulation.

Based on Eqs. (3.33) and (3.34), the values of R_1 and M_1 under different duty cycles can be calculated. Table 3.5 shows the relationship between M_1 and the duty cycle for adaptive M-QAM with dimming control. As expected, the value of M_1 increases as the duty cycle is reduced. Figure 3.14 (a) shows the relationship between the adaptive

Table 3.5 Relationship between M_1 and duty cycle for adaptive M-QAM with dimming control.

Duty cycle	0.1	0.2	0.3	0.4	0.5	0.6	0.7	0.8	0.9	1.0
M_1	256	128	64	32	16	16	8	8	8	4

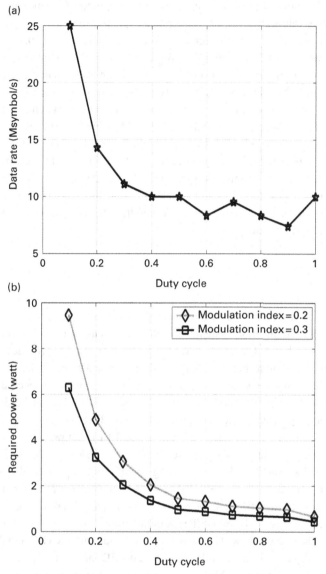

Figure 3.14 (a) Adaptive symbol rate of M-QAM signal versus the duty cycle; (b) the required LED lamp power to achieve BER of 10^{-3} without applying dimming control versus the duty cycle.

symbol rate R_1 and the duty cycle. When the duty cycle is less than 0.3, the adaptive symbol rate has to increase significantly to satisfy the communication quality. The highest adaptive rate is 2.5 times as high as the original symbol rate when the duty cycle is 0.1, which is the same as that of an OOK signal when the duty cycle is 0.4, as shown in Fig. 3.13 (a). Hence, the increase in the symbol rate of an M-QAM OFDM signal is moderate compared with that of an OOK signal.

Figure 3.14 (b) shows the relationship between the required LED lamp power and the duty cycle. When the duty cycle is greater than 0.4, the required LED lamp power increases slowly as the duty cycle decreases. When the duty cycle is 0.9, the required LED lamp powers without applying dimming control are 0.97 W and 0.65 W for the modulation indices of 0.2 and 0.3, respectively, which are larger than the required LED lamp power of an OOK signal with a 0.1 duty cycle. However, the required LED lamp power increases rapidly when the duty cycle is reduced below 0.4. When the duty cycle is 0.1, the required LED lamp powers of M-QAM OFDM signals are 9.5 W and 6.3 W for the 0.2 and 0.3 modulation indices, respectively, which are more than 13 times larger than the corresponding required LED lamp powers of an OOK signal. Hence, when the duty cycle is larger than 0.4, the adaptive M-QAM OFDM signal remains to be a good choice to be incorporated with a dimming control scheme, since the required LED lamp power without applying dimming control is about 2 W and the adaptive symbol rate is not larger than the original symbol rate.

3.5   Summary

This chapter describes three recently developed approaches to improve the performance of visible light communication systems. Section 3.2 discusses a receiver plane tilting technique to reduce the SNR fluctuation in a room. This scheme can provide 5.69 dB and 4.14 dB improvement for the peak-to-trough SNR performance, for the cases of one LED lamp and four LED lamps, respectively. The corresponding maximum spectral efficiency improvements are 0.47 bit/s/Hz and 0.23 bit/s/Hz, respectively. Section 3.3 describes an LED lamp arrangement technique to improve SNR and BER performances. It is shown that the arrangement of 12 LED lamps in a circle and 4 LED lamps at corners can attain much better performance than other possible arrangements. By applying the time domain zero-forcing equalization, this LED lamp arrangement is able to provide almost identical communication qualities in terms of SNR and BER to all users at different positions throughout the room. In Section 3.4, the performance of a VLC system under a dimming control scheme is discussed for two different modulation formats, namely OOK and M-QAM OFDM. The results show that the adaptive data rate of an OOK signal is always larger than the original data rate, while the required LED lamp power is less than 1 W, when the original data rate is 10 Mbit/s. The adaptive data rate of the M-QAM OFDM signal is not larger than the original symbol rate when the duty cycle is larger than 0.4, however, the LED lamp required without applying dimming control is always larger than that required by the OOK signal.

References

[1] T. Komine and M. Nakagawa, "Fundamental analysis for visible-light communication system using LED lights," *IEEE Transactions on Consumer Electronics*, **50**, 100–107, 2004.

[2] D. C. O'Brien, G. Faulkner, K. Jim, *et al.*, "High-speed integrated transceivers for optical wireless," *IEEE Communications Magazine*, **41**, 58–62, 2003.

[3] J. Vucic, C. Kottke, K. Habel, and K. D. Langer, "803 Mbit/s visible light WDM link based on DMT modulation of a single RGB LED luminary," in *Optical Fiber Communication/*National Fiber Optic Engineers Conference (OFC/NFOEC), 2011, pp. 1–3.

[4] H. Elgala, R. Mesleh, and H. Haas, "Indoor broadcasting via white LEDs and OFDM," *IEEE Transactions on Consumer Electronics*, **55**, 1127–1134, 2009.

[5] J. M. Kahn and J. R. Barry, "Wireless infrared communications," *IEEE Proceedings*, **85**, 265–298, 1997.

[6] K. D. Langer and J. Grubor, "Recent developments in optical wireless communications using infrared and visible light," in International Conference on *Transparent Optical Networks (ICTON)*, 2007, pp. 146–151.

[7] S. Hann, J.-H. Kim, S.-Y. Jung, and C.-S. Park, "White LED ceiling lights positioning systems for optical wireless indoor applications," in European Conference and Exhibition on *Optical Communication (ECOC)*, 2010, pp. 1–3.

[8] L. Zeng, D. O'Brien, L.-M. Hoa, *et al.*, "Improvement of data rate by using equalization in an indoor visible light communication system," in International Conference on *Circuits and Systems for Communications (ICCSC)*, 2008, pp. 678–682.

[9] G. Ntogari, T. Kamalakis, and T. Sphicopoulos, "Performance analysis of space time block coding techniques for indoor optical wireless systems," *IEEE Journal on Selected Areas in Communications*, **27**, 1545–1552, 2009.

[10] D. Bykhovsky and S. Arnon, "An experimental comparison of different bit-and-power-allocation algorithms for DCO-OFDM," *Journal of Lightwave Technology*, **32**, 1559–1564, 2014.

[11] D. Bykhovsky and S. Arnon, "Multiple access resource allocation in visible light communication systems," *Journal of Lightwave Technology*, **32**, 1594–1600, 2014.

[12] G. Cossu, A. M. Khalid, P. Choudhury, R. Corsini, and E. Ciaramella, "3.4 Gbit/s visible optical wireless transmission based on RGB LED," *Optics Express*, **20**, B501–B506, 2012.

[13] D. Tsonev, H. Chun, S. Rajbhandari, *et al.*, "A 3-Gb/s single-LED OFDM-based wireless VLC link using a gallium nitride μLED," *IEEE Photonics Technology Letters*, **26**, 637–640, 2014.

[14] T. Fath, M. Di Renzo, and H. Haas, "On the performance of space shift keying for optical wireless communications," in *IEEE GLOBECOM Workshops (GC Wkshps)*, 2010, pp. 990–994.

[15] R. Mesleh, R. Mehmood, H. Elgala, and H. Haas, "Indoor MIMO optical wireless communication using spatial modulation," in IEEE International Conference on *Communications (ICC)*, 2010, pp. 1–5.

[16] R. Mesleh, H. Elgala, and H. Haas, "Optical spatial modulation," *IEEE/OSA Journal of Optical Communications and Networking*, **3**, 234–244, 2011.

[17] T. Fath and H. Haas, "Performance comparison of MIMO techniques for optical wireless communications in indoor environments," *IEEE Transactions on Communications*, **61**, 733–742, 2013.

[18] http://www.ted.com/talks/harald_haas_wireless_data_from_every_light_bulb.html.

[19] J. Chen, C. Yu, Z. Wang, J. Shen, and Y. Li, "Indoor optical wireless integrated with white LED lighting: Perspective & challenge," in 10th International Conference on *Optical Communications and Networks (ICOCN)*, 2011, pp. 1–2.

[20] Z. Wang, W. D. Zhong, C. Yu, *et al.*, "Performance of dimming control scheme in visible light communication system," *Optics Express*, **20**, 18861–18868, 2012.

[21] Z. Wang, C. Yu, W. D. Zhong, and J. Chen, "Performance improvement by tilting receiver plane in M-QAM OFDM visible light communications," *Optics Express*, **19**, 13418–13427, 2011.

[22] Z. Wang, C. Yu, W. D. Zhong, J. Chen, and W. Chen, "Performance of a novel LED lamp arrangement to reduce SNR fluctuation for multi-user visible light communication systems," *Optics Express*, **20**, 4564–4573, 2012.

[23] Z. Wang, W. D. Zhong, C. Yu, and J. Chen, "A novel LED arrangement to reduce SNR fluctuation for multi-users in visible light communication systems," in 8th International Conference on *Information, Communications and Signal Processing (ICICS)*, 2011, pp. 1–4.

[24] Z. Wang, C. Yu, W. D. Zhong, J. Chen, and W. Chen, "Performance of variable M-QAM OFDM visible light communication system with dimming control," in 17th *Opto-Electronics and Communications* Conference *(OECC)*, 2012, pp. 741–742.

[25] Z. Wang, J. Chen, W. D. Zhong, C. Yu, and W. Chen, "User-oriented visible light communication system with dimming control scheme," in 11th International Conference on *Optical Communications and Networks (ICOCN)*, 2012, pp. 1–4.

[26] H. Kressel, *Semiconductor Devices for Optical Communication*, Springer-Verlag, 1982.

[27] J. R. Barry, *Wireless Infrared Communications*, Kluwer Academic Publishers, 2006.

[28] I. Neokosmidis, T. Kamalakis, J. W. Walewski, B. Inan, and T. Sphicopoulos, "Impact of nonlinear LED transfer function on discrete multitone modulation: Analytical approach," *IEEE Journal of Lightwave Technology*, **27**, 4970–4978, 2009.

[29] C. H. Edwards and D. E. Penney, *Calculus*, Prentice Hall, 2002.

[30] M. T. Heath, *Scientific Computing – An Introductory Survey*, McGraw-Hill, 2002.

[31] A. Svensson, "An introduction to adaptive QAM modulation schemes for known and predicted channels," *IEEE Proceedings*, **95**, 2322–2336, 2007.

[32] J. Proakis, *Digital Communications*, 3rd ed., McGraw-Hill, 1995.

[33] F. Xiong, *Digital Modulation Techniques*, 2nd ed., Artech House, 2006.

[34] R. J. Essiambre, G. Kramer, P. J. Winzer, G. J. Foschini, and B. Goebel, "Capacity limits of optical fiber networks," *IEEE Journal of Lightwave Technology*, **28**, 662–701, 2010.

[35] M. Z. Afgani, H. Haas, H. Elgala, and D. Knipp, "Visible light communication using OFDM," in International Conference on *Testbeds and Research Infrastructures for the Development of Networks and Communities (TRIDENTCOM)*, 2006, pp. 6–134.

[36] J. Armstrong, "OFDM for optical communications," *IEEE Journal of Lightwave Technology*, **27**, 189–204, 2009.

[37] U. S. Jha and R. Prasad, *OFDM towards Fixed and Mobile Broadband Wireless Access*, Artech House, 2007.

[38] A. Goldsmith, *Wireless Communications*, Cambridge University Press, 2005.

[39] H. Sugiyama, S. Haruyama, and M. Nakagawa, "Brightness control methods for illumination and visible-light communication systems," in 3rd International Conference on *Wireless and Mobile Communications (ICWMC)*, 2007, pp. 78–83.

[40] J.-H. Choi, E.-B. Cho, T.-G. Kang, and C. G. Lee, "Pulse width modulation based signal format for visible light communications," in 15th *Opto-Electronics and Communications* Conference *(OECC)*, 2010, pp. 276–277.

[41] S. Rajagopal, R. D. Roberts, and S.-K. Lim, "IEEE 802.15.7 visible light communication: Modulation schemes and dimming support," *IEEE Communications Magazine*, **50**, 72–82, 2012.

[42] G. Ntogari, T. Kamalakis, J. Walewski, and T. Sphicopoulos, "Combining illumination dimming based on pulse-width modulation with visible-light communications based on discrete multitone," *IEEE/OSA Journal of Optical Communications and Networking*, **3**, 56–65, 2011.

[43] J. Proakis and M. Salehi, *Contemporary Communication Systems Using MATLAB*, PWS Pub., 1998.

[44] A. J. Goldsmith and S.-G. Chua, "Variable-rate variable-power MQAM for fading channels," *IEEE Transactions on Communications*, **45**, 1218–1230, 1997.

4   Light positioning system (LPS)

Mohsen Kavehrad and Weizhi Zhang

One of the most promising applications of visible light communication is indoor positioning, also referred to as indoor localization. Indoor positioning has many applications in real life. For instance, the development of this technique will make possible accurate location-based services (LBS), which have increased dramatically together with the popularization of mobile computing devices such as smart phones and tablets. Therefore, it is increasingly attracting researchers' interest. Current research on visible light communication (VLC) technology has provided a new approach to commercial high-quality indoor positioning systems. Through this chapter, we first list all the applications, investigate current access to the radio spectrum, and cover the merits of shifting the needs of indoor positioning to the visible light spectrum. A survey of VLC-based indoor positioning techniques and related works follows. Finally, challenges and potential solutions are discussed.

4.1   Indoor positioning and merits of using light

Indoor positioning is a key technology which can be beneficial for many industries and customers (Fig. 4.1). These include location sensing and management of products inside large warehouses, indoor navigation services for pedestrians inside large buildings such as museums and shopping malls, location-based services (LBS) and advertisements for consumers etc.

According to a report by the Federal Communications Commission in 2012 [1], most research agrees that the location-based services market will be tripled by 2015 compared to its size in 2012. The report also stated that Foursquare, a location-based social networking company, had 10 million users, but this has been increasing since then and was 40 million in 2013 [2]. Research also predicts that the total mobile LBS market will be worth more than 10 billion US dollars in 2015 [3].

What is maybe even more important, with the standardization of femto-cells in the mobile network, is that location awareness of users will be critical to address hand-off and resource allocation issues, which are discussed next.

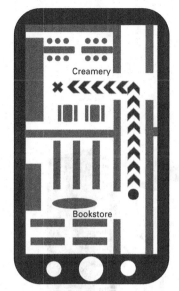 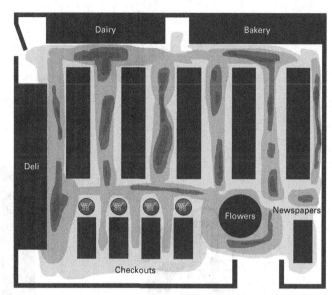

Figure 4.1   (a) Indoor navigation for pedestrians; (b) location analytics based on indoor positioning.

4.1.1   Introduction to indoor positioning

Although the Global Positioning System (GPS) is well-developed and performs well for outdoor applications, it is still very difficult to use it indoors (see Fig. 4.2) because of the poor coverage of satellite signals. The main reason behind this is the multipath effects of radio-waves on the GPS signals.

Signals emitted from satellites may reflect off surfaces around the receiver, such as trees, roofs, walls, or even human bodies. The signals subject to multipath will arrive at the receive sensor as delayed components, therefore creating positioning errors. For an indoor environment, this induced inaccuracy is so large that the performance of GPS is downgraded to an unacceptable level. Apart from multipath, there are potentially artificial sources of interference, whether intentional or not, further bringing down the positioning accuracy. As a result of the poor performance of the GPS system, currently commercialization of indoor positioning systems is still in its infancy and there is a need to develop a reliable and accurate system. So far, different proposed candidate systems try to perform accurate indoor positioning by utilizing radio-wave, acoustic and optical technologies [4–9].

4.1.2   Spectrum crunch and future mobile system

With the increasing popularity of multimedia services supplied over the cellular radio frequency (RF) networks [10], including data services such as web browsing, audio and video on demand, it is certainly only a matter of time before users will face extreme congestion while trying to connect to avail themselves of these services. Advances in displays, battery technology and processing power have made it possible for users to

Figure 4.2   Indoor coverage problems of Global Positioning System (GPS).

afford and carry around smart phones and tablets. Hence, as we are entering a new era of always-on connectivity, the expectation from users for not only ubiquitous but also seamless voice and video services presents a significant challenge for today's telecommunications systems. The prospects for the delivery of such multimedia services to these users are crucially dependent on the development of low-cost physical layer delivery mechanisms [11]. It is recognised that the electromagnetic spectrum has become extremely crowded [12]. For the benefit of obtaining more bandwidth, the frequency at which mobile bands are regulated has already shifted from 450~800 MHz in the first generation of cellular systems, to as high as 2.6~2.7 GHz in today's 4G technology.

As frequency increases, the path loss increases, perhaps as frequency squared to frequency to some higher power, depending on environment. There is no cliff as frequency increases; it just becomes more difficult gradually to deliver a high quality-of-service over a non-line-of-sight (NLOS) path. This issue becomes less important as the base station spacing has decreased to increase system capacity and that is the best way to do it. Decrease spacing by a factor of two and capacity increases by a factor of four, i.e. two squared. Close-spaced base stations need less link margin to overcome path loss so the increased path loss with higher frequency tends to be offset by the closer spacing. Allocating more bandwidth to cellular would help a little with the spectrum crowding, but probably not enough to take care of the major issue. Even if you had twice the bandwidth available, that is a small help.

Therefore, decreasing base station spacing (increasing base station density) is a better way to provide more capacity and that is what has been done for the last 20 years. The

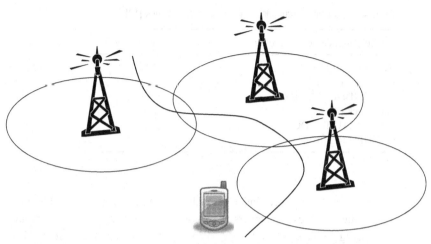

Figure 4.3   Handover prediction in mobile networks.

wireless industry is talking about femto-cells for a high capacity backhauling. However, smaller cells can bring problems that were not noticeable before, e.g. more frequent handovers, more complex resource allocation and re-use, etc. Precise indoor positioning techniques will benefit the overall network management since femto-cells are designed to serve indoor environments. By using location information from users the base station can make better frequency allocation to virtually provide more available bandwidth to users, as well as better prediction about when handover will happen (Fig. 4.3), to realize a seamless service.

4.1.3   Advantages of VLC-based positioning

As noted in the previous chapters, VLC technology has great potential and will help offload traffic from the highly congested radio-wave band. First, the visible light band theoretically provides 400 THz bandwidth (375–780 nm), which is a much larger bandwidth than RF techniques can utilize. Moreover, lightwaves in the optical wavelength range are confined to the walls in a room and generally do not penetrate solid materials. Hence, practical and usable networks can be readily realized that utilize this self-limiting link distance. We call such systems that employ this property high-bandwidth islands. The motivation for operators to actually choose to transfer data through this optical band is that by doing so, the entire huge bandwidth can be re-used next door, free of interference. Optical wireless is a future proof solution since additional capacity far beyond the capabilities of radio could be delivered to users as their needs increase with time.

Light emitting diodes (LEDs) will replace today's incandescent and fluorescent lamps used for lighting in due course. Compared to traditional light devices, LED has higher power efficiency, longer lifetime and higher tolerance to environmental hazards. Since LED is a semiconductor light source, it can be easily modulated for many applications

Table 4.1 Positioning accuracies of radio-wave techniques [13].

Positioning method (technology)	Accuracy (m)
Sapphire Dart (UWB)	0.3
Ekahau (WLAN)	1
TOPAZ (Bluetooth+IR)	2
SnapTrack (AGPS)	5–50
WhereNet (UHF TDOA)	2–3
LANDMARC (RFID)	2

other than lighting, such as indoor broadband communication, smart lighting, etc. [11, 12]. This feature enables researchers to explore the possibility to use it to address the indoor positioning problem. Recently, VLC using LEDs or other light sources has been considered as one of the most attractive solutions for indoor positioning systems because of the many features it brings, as described below:

Better positioning accuracy

There are many proposed indoor positioning solutions based on radio-wave techniques. Wireless technologies associated with them include, but are not limited to: wireless local area network (WLAN), radio-frequency identification (RFID), cellular, ultra-wide band (UWB), Bluetooth, etc. [4–7, 21]. These methods deliver positioning accuracies from tens of centimeters to several meters [13], see Table 4.1. VLC-based systems are expected to provide better positioning accuracy than radio-wave solutions, since they suffer less from multipath effects and interference from other wireless handsets, which will be verified in the following sections.

Generate no radio-frequency (RF) interference

Aside from the relatively lower accuracies radio-wave approaches can provide, the RF electro-magnetic (EM) interference brought in is always a concern for many indoor environments. On the one hand, EM radiation generated by these techniques will occupy the already-congested limited mobile band, further degrading the performance of other wireless devices. On the other hand, since radio-frequency interference can disable certain kinds of medical devices, with results ranging from inconvenience to accidental injuries, radio-frequency radiation is restricted, or even prohibited in hospitals, as well as in many other places having interference concerns.

In contrast, VLC systems for communication or positioning purposes, do not generate any RF interference and therefore are safe to use inside healthcare facilities. LEDs can be used to carry different forms of information, e.g. biomedical information from monitoring devices, text information from patients, etc. to medical staff [14], [15].

Re-use current lighting infrastructures

Positioning techniques based on ultrasound and other acoustic waves offer accuracy up to several centimeters [13], however, they require a dense and calibrated grid of transmitters, which sometimes may dramatically increase the cost of the system. In contrast, VLC-based systems re-use the current light infrastructures and thus widespread coverage is assured. At the same time, little or no renovation is needed to provide a service, therefore VLC-based systems also offer very economical solutions for indoor positioning requirements.

4.2 Positioning algorithms

Investigations within this field demonstrate that positioning algorithms proposed so far can be categorized into three types: triangulation, scene analysis, and proximity.

4.2.1 Triangulation

Triangulation is the general name for positioning algorithms which use the geometric properties of triangles for location estimation. Triangulation has two branches: lateration and angulation [13]. In lateration methods, the target location is estimated by measuring its distances from multiple reference points. For all the VLC-based indoor positioning systems proposed, the reference points are light sources and the target is an optical receiver. The distances are almost impossible to measure directly; however, they can be mathematically calculated from other measurements such as received signal strength (RSS), time-of-arrival (TOA) or time-difference-of-arrival (TDOA). On the other hand, angulation measures angles relative to several reference points (angle-of-arrival (AOA)), after which location estimation is carried out by finding intersection points of direction lines which are radii from reference points.

4.2.2 Triangulation – circular lateration

Circular lateration methods mainly make use of two kinds of measurements: TOA or RSS.

As light travels at a constant speed in air, the distance between the receiver and light source is proportional to the travel time of the optical signals. In TOA-based systems, time-of-arrival measurements with respect to three light sources are required to locate the target, giving intersection of three circles for 2-D, or three spheres in a 3-D scenario. A very good example of TOA-based systems is the widely used GPS system. In the system, navigation messages sent from satellites contain time information (in the form of a ranging code) and Ephemeris (orbit information for all the satellites). After successfully receiving navigation messages from more than three satellites, circular lateration (referred to as trilateration) is performed to determine the receiver's location. However, all the clocks used by the transmitters as well as by the receiver have to be perfectly synchronized. For indoor applications, the positioning accuracy should range from sub-meter to centimeters, which means the different clocks in TOA-based systems have to be synchronized at the level of a few nanoseconds, or even higher accuracy. As a result, the complexity and cost of such systems are impracticable. Therefore, research on the TOA-based positioning technique has been very limited. Cramer–Rao bound analysis of a

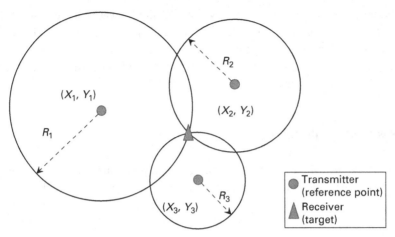

Figure 4.4   Positioning using circular lateration.

TOA-based VLC positioning system considering only shot noise is given in [16], and the results show that accuracy around 2 ~ 5 cm can be achieved, depending on different system settings.

RSS-based systems calculate the propagation loss the emitted signal has experienced based on measurements of received signal strength. Range estimation is then made by using a proper path loss model. As in TOA-based systems, estimation of the target's position is obtained by circular lateration, shown in Fig. 4.4. Because of the availability of line-of-sight (LOS) channels for most indoor environments, it is considered that RSS-based methods using VLC will deliver a good performance. According to simulations performed in previous work [17], the target can be located with a positioning error around 0.5 mm. In [18], taking the rotation and moving speed of the receiver into consideration, the authors show that an overall accuracy around 2.5 cm can be obtained, assuming a typical moving speed of the receiver. In [19], the authors account for possible installation errors of the LED lighting bulbs as well as the orientation angle of the target and show that a precision of 5.9 cm with 95% confidence can be expected, given indirect sunlight exposure and proper installation of the LED bulbs.

Now let us introduce a mathematical expression for circular lateration in two-dimensional space. The expression for three-dimensions is similar. Let (X_i, Y_i) represent the position of the ith transmitter (reference point) on a two-dimensional plane and (x, y) denote the position of the receiver (target). If the measured distance between the ith transmitter and receiver is R_i, then every circle as shown in Fig. 4.4 is a set of possible locations of the receiver determined by a single range measurement, which is:

$$(X_i - x)^2 + (Y_i - y)^2 = R_i^2, \tag{4.1}$$

where $i = 1, 2, \ldots, n$ and n is the number of transmitters (reference points) involved in range measurements. Theoretically, if range measurements are noise-free, the intersection of circles given by Eq. (4.1) should yield the position of the receiver as a single point.

However, this cannot be realistic in real measurements. Noisy range measurements lead to multiple solutions to our system described by Eq. (4.1). In this scenario, the least squares solution, which is discussed in [20, 21], provides a standard approach to the approximate solution of this positioning system.

Note that:

$$R_i^2 - R_1^2 = (x - X_i)^2 + (y - Y_i)^2 - (x - X_1)^2 - (y - Y_1)^2 \qquad (4.2)$$

$$= X_i^2 + Y_i^2 - X_1^2 - Y_1^2 - 2x(X_i - X_1) - 2y(Y_i - Y_1), \qquad (4.3)$$

where $i = 1, 2, \ldots, n$. Then we can rewrite the equations describing the system into matrix form [21]:

$$AX = B, \qquad (4.4)$$

where

$$X = [x\ y]^T \qquad (4.5)$$

$$A = \begin{bmatrix} X_2 - X_1 & Y_2 - Y_1 \\ \vdots & \vdots \\ X_n - X_1 & Y_n - Y_1 \end{bmatrix} \qquad (4.6)$$

and

$$B = \frac{1}{2} \begin{bmatrix} (R_1^2 - R_2^2) + (X_2^2 + Y_2^2) - (X_1^2 + Y_1^2) \\ \vdots \\ (R_1^2 - R_n^2) + (X_n^2 + Y_n^2) - (X_1^2 + Y_1^2) \end{bmatrix}. \qquad (4.7)$$

Then the least square solution to the system is given by:

$$X = (A^T A)^{-1} A^T B. \qquad (4.8)$$

4.2.3 Triangulation – hyperbolic lateration

Hyperbolic lateration methods are usually associated with TDOA measurements. In TDOA-based systems, light signals from different LEDs are designed to be transmitted at the same instant. This can be easily achieved since all the LEDs are in close proximity so they can share the same clock. The receiver measures the difference in time at which these signals arrive. On the other hand, the receiver does not need to be synchronized with the transmitters, since it is not trying to extract the absolute time of arrival information.

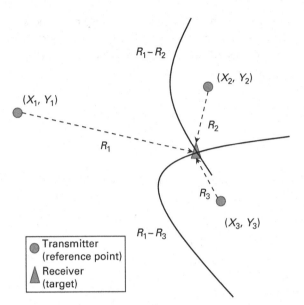

Figure 4.5   Positioning using hyperbolic lateration.

As in TOA and RSS based systems, three light sources are needed to enable 2-D or 3-D positioning. Because a single TDOA measurement with two light sources involved provides a hyperboloid on a 2-D plane or a hyperbola in a 3-D space, two TDOA measurements are required to locate the target by using hyperbolic lateration (referred to as multilateration). Instead of taking TDOA measurements directly, we may take other measurements, and compute TDOA information by using them. In [22], sinusoidal components of signals emitted from two LEDs generate an interference pattern at the receiver because they have the same frequency. Thus, the sinusoidal peak-to-peak value of the received signal can be utilized to get equivalent TDOA measurements. In [23], TDOA information is obtained by detecting phase differences between three signals with different frequencies. Computer simulations show an overall positioning accuracy of 1.8 mm.

For mathematical expression of hyperbolic lateration in two-dimensional space, following the notation used for circular lateration, every hyperbola as shown in Fig. 4.5 is a set of possible locations of the receiver determined by a single measurement of range difference. Every hyperbola can be represented by:

$$D_{ij} = R_i - R_j = \sqrt{(X_i - x)^2 + (Y_i - y)^2} - \sqrt{(X_j - x)^2 + (Y_j - y)^2}, \qquad (4.9)$$

where D_{ij} denotes the difference between the ranges R_i and R_j, with respect to the ith and jth reference points and $i{\neq}j$. Note that:

$$(R_1 + D_{i1})^2 = R_i^2, \qquad (4.10)$$

$$X_i^2 + Y_i^2 - X_1^2 - Y_1^2 - 2x(X_i - X_1) - 2y(Y_i - Y_1) - D_{i1} - 2D_{i1}R_1 = 0, \qquad (4.11)$$

where $i = 1, 2, \ldots, n$. We can then rewrite the equations describing the system into matrix form [21]:

$$AX = B, \tag{4.12}$$

where

$$X = [x\ y\ R_1]^T, \tag{4.13}$$

$$A = \begin{bmatrix} X_2 - X_1 & Y_2 - Y_1 & D_{21} \\ \vdots & \vdots & \vdots \\ X_n - X_1 & Y_n - Y_1 & D_{n1} \end{bmatrix}, \tag{4.14}$$

and

$$B = \frac{1}{2} \begin{bmatrix} (X_2^2 + Y_2^2) - (X_1^2 + Y_1^2) - D_{21}^2 \\ \vdots \\ (X_n^2 + Y_n^2) - (X_1^2 + Y_1^2) - D_{n1}^2 \end{bmatrix}. \tag{4.15}$$

Then the least squares solution to the system is given by:

$$X = (A^T A)^{-1} A^T B. \tag{4.16}$$

4.2.4   Triangulation – angulation

In AOA-based systems, the receiver measures angles of arriving signals from several reference points. The target's location is then determined as the intersection of direction lines (Fig. 4.6). Typically two light sources are needed to realize 2-D and three for 3-D

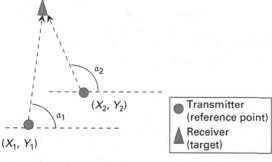

Figure 4.6   Positioning using angulation.

positioning. Interestingly, we can find a historical trace back to older photogrammetry techniques [24] with many similarities. The greatest advantage of AOA-based systems is that no time synchronization is needed. Another advantage is that it is relatively easier to detect the AOA of the incoming signals in the optical domain with an imaging receiver, compared to employing complex antenna arrays used in radio-wave approaches. The widely deployed front-facing cameras, which are inherently imaging receivers, on smart phones and tablets have brought opportunities for this method to become practicable on mobile consumer electronics. However, to achieve a good performance for a real system, the lighting infrastructure may have to be adjusted since most of these cameras have a very limited field-of-view (FOV). More generally, the positioning accuracy of an AOA-based system will downgrade when the target gets farther from the light sources due to the limited spatial resolution of the imaging receivers. A positioning accuracy of 5 cm is reported when using an imaging receiver with a resolution of 1296×964 pixels [25]. In [26], researchers proposed a two phase positioning algorithm which makes use of both RSS and AOA measurements, taking multipath effects due to reflections into consideration. Computer simulations showed that a median accuracy of 13.95 cm can be achieved.

To obtain the least square solution for an AOA-based system, suppose α_i denotes the measured angle-of-arrival with respect to the ith transmitter, which is given by:

$$\tan \alpha_i = \frac{y - Y_i}{x - X_i}, \tag{4.17}$$

where $i = 1, 2, \ldots, n$. Then we will have,

$$(x - X_i) \sin \alpha_i = (y - Y_i) \cos \alpha_i. \tag{4.18}$$

Then we can rewrite the equations describing the system into matrix form [21]:

$$AX = B, \tag{4.19}$$

where

$$X = [x \, y]^T, \tag{4.20}$$

$$A = \begin{bmatrix} -\sin \alpha_1 & \cos \alpha_1 \\ \vdots & \vdots \\ -\sin \alpha_n & \cos \alpha_n \end{bmatrix}, \tag{4.21}$$

and

$$B = \begin{bmatrix} Y_1 \cos \alpha_1 - X_1 \sin \alpha_1 \\ \vdots \\ Y_n \cos \alpha_n - X_n \sin \alpha_n \end{bmatrix}. \tag{4.22}$$

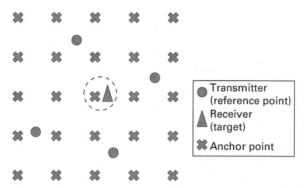

Figure 4.7   Positioning system using scene analysis.

Then the least square solution to the system is given by:

$$X = (A^T A)^{-1} A^T B. \tag{4.23}$$

4.2.5   Scene analysis

Scene analysis refers to positioning algorithms which make use of fingerprints associated with every anchor point in the system inside a scene, as shown in Fig. 4.7. The target is then located by matching real-time (on-line) measurements to these fingerprints. Measurements that can be used as fingerprints include all measurements mentioned earlier, i.e. TOA, TDOA, RSS and AOA. RSS is the most used form of fingerprint. The time required to match fingerprints is usually shorter than performing a triangulation, thus saving a lot of time and power which will otherwise be used for computing. However, scene analysis solutions also have a significant disadvantage. They cannot be deployed instantly inside a new scenario since accurate system pre-calibration is needed. A scene analysis method utilizing RSS information from four LED lights as fingerprints shows that a positioning accuracy of 4.38 cm can be attained [27].

4.2.6   Proximity

Proximity-based systems (Fig. 4.8) rely on a dense grid of light sources, each having a known position and a unique hardware ID. When the target receives signals from a single light source, it is considered to be co-located with the source. When signals from multiple sources are detected, averaging will be performed. Proximity systems using VLC technology theoretically provide accuracies no higher than the resolution of the lighting grid itself. Note that when dense grids are employed, a narrow illumination beam from the light sources is required in order to prevent interference and impaired location determination. In [9], the authors experimentally demonstrated a proximity-based indoor positioning system. The positioning is provided by visible light LEDs, while a ZigBee

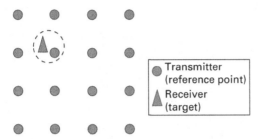

Figure 4.8   Proximity positioning system.

wireless network is employed to send the location information to the main node, extending the working range of the system.

4.2.7   Comparison of positioning techniques

To compare the systems we have mentioned, we adopt the following performance metrics: accuracy, spatial dimension, and complexity, see Table 4.2.

Accuracy is usually referred to as the mean value of positioning error. Here, the positioning error is defined as the Euclidean distance between the real location of the target and the estimation of it. The smaller the mean value is, the higher accuracy a system can achieve. So naturally this is the most important factor when we evaluate a positioning system.

Spatial dimension is the number of dimensions a positioning system is able to provide on location information. Many proposed solutions are only capable of 2-D positioning horizontally, in which case if the height of the target changes, the positioning accuracy of the system will drop as a result. Systems capable of 3-D positioning, on the other hand, provide a better performance in this scenario.

The complexity defined in this chapter contains two components. The first is the system requirement on hardware, i.e. how many devices are involved and how complicated is the overall system configuration. Hardware complexity mainly determines what would be the deployment cost of an indoor positioning system. The other component is the computational complexity of the positioning algorithm, or the delay before the system can update the current location of the target. In most of the systems, data processing happens on the target side, e.g. mobile handsets in real scenarios. Algorithm complexity is also important, because battery life is still a huge concern for handsets, even though they are currently capable of powerful computing.

4.3   Challenges and solutions

4.3.1   Multipath reflections

As mentioned earlier, a path loss model is employed in an RSS-based system. Current optical path loss models, which only take the LOS propagation into consideration, do not

Table 4.2 Comparison of solutions mentioned.

Reference no.	Positioning algorithm	Accuracy	Space dimension	Complexity Hardware/ Algorithm	Comments
17	RSS	0.5 mm (Simulation)	2-D/3-D	Moderate/ Moderate	3-D positioning is realized
18	RSS	2.5 cm (Simulation)	2-D	Moderate/ Moderate	Rotation and moving speed of the receiver are accounted for
19	RSS	5.9 cm (Simulation)	2-D	Low/Moderate	Asynchronous system design, installation errors of LED bulbs and orientation angle of the target are taken into consideration
23	TDOA	1.8 mm (Simulation)	2-D	High/Moderate	Frequency-division-multi-access (FDMA) protocol employed
25	AOA	4.6 cm (Experiment)	3-D	High/Moderate	Different colors are used to distinguish LED lights
26	RSS+AOA	13.95 cm (Simulation, median value)	2-D/3-D	High/High	High Lambertian order sources are assumed ($m = 30$), multipath effects and receiver orientation are taken into consideration
27	Scene analysis (RSS based)	4.38 cm (Experiment, mean value)	2-D	Moderate/Low	Pre-calibration needed
9	Proximity	Room level (Experiment)	2-D	Low/Low	4 MHz carrier is employed to achieve better optical detection

always hold due to multipath effects caused by reflections of light on different surfaces (e.g. walls). The optical energy contained within these reflected components would be directly translated into extra positioning errors, thus positioning accuracy will be downgraded.

One available solution is to use a fly-eye receiver [28]. As an imaging receiver, it is able to resolve light components coming from different directions. Therefore, the effects of reflections can be significantly reduced by necessary signal processing after optical detection. Furthermore, the receiver can provide angle information which can be used to improve the positioning accuracy in combination with AOA positioning.

4.3.2   Synchronization

In the systems based on time measurements (TOA, TDOA), synchronization is a major source of positioning error. In TOA-based systems, it is extremely difficult to get all transmitters and the receiver precisely synchronized, if not impossible. This becomes easier in TDOA-based approaches. The clock used by the receiver does not have to be precisely synchronized with the transmitters and we can make use of other measurements as mentioned in [19, 20]. Nevertheless, the signals from all transmitters have to be emitted simultaneously. Otherwise, the initial phases of transmitted signals must be measured in a very accurate way. Therefore, synchronization has to be performed between transmitters, which can lead to potentially high system deployment costs and limited applicable scenarios.

4.3.3   Channel multi-access

As implied by the principles of the positioning algorithms mentioned above, in all systems except proximity-based ones, multiple transmitters are required to estimate the receiver position. This means we have to solve the channel multi-access problem to avoid interference. Even for proximity-based systems, the problem has to be handled to minimize the possibility of detection failure when multiple transmitted signals fall in the receiver field-of-view, if no channel multi-access protocol is employed and transmitters keep sending out signals all the time.

In GPS systems, channel multi-access is addressed by using the code-division-multi-access (CDMA) technique [29]. In a TDOA-based system proposed in [23], frequency-division-multi-access (FDMA) is selected as a solution. Time-division-multi-access (TDMA) is another technique used by many systems; it requires synchronization between transmitters, resulting in higher deployment cost and time. In [19], a new protocol is proposed to eliminate the need for synchronization. The adoption of the framed slotted ALOHA protocol successfully enables the LEDs to work asynchronously, thus no physical connections between them are required, lowering system complexity and cost.

4.3.4   Service outage

One very realistic problem with light positioning systems can be expressed in the form of a simple question: What will happen when there is no light? Indeed this is a serious concern since there might be occasions where users demand a positioning service without the lights being turned on. Also, light can be blocked by furniture and human bodies, resulting in service outage. Therefore, this problem must be addressed to ensure that users will have access to the positioning service most of the time.

Two approaches can offer help regarding this issue. First, infrared laser or LED can be integrated into the LED bulbs used in the future positioning system, to continue offering a service when visible light illumination is no longer desired. Secondly, to eliminate the temporary service outage mainly due to obstruction, sensor fusion technology [30] making use of the inertial sensors inside the handset can be employed. It will help the system to give the best performance by integrating inertial navigation, which will guide the user through the obstruction period, and optical positioning, which generally gives more accurate estimation.

4.3.5 Privacy

Although it does not seem a technical issue, privacy is a huge concern for light positioning systems. Indeed, better positioning accuracy makes people more worried about the security of the real-time location information. In the visionary report from the FCC [1], the committee identified several privacy issues as follows: notice and transparency, meaningful consumer choice, third party access to personal information, and data security and minimization. So we suggest that research on access control and encryption is necessary for the long-term development of light positioning systems.

4.4 Summary

Indoor positioning is one of the most important and promising applications associated with visible light communications technology. We have specifically discussed why accurate indoor positioning plays a key role in future mobile networks, considering the fact that the mobile band is currently heavily congested. Optical positioning systems are expected to provide better accuracy compared to radio-wave rivals. They will not generate radio-frequency (RF) interference, maintaining the current performance of other wireless handsets as well as making these suitable to be deployed inside RF restricted or prohibited environments. Finally, indoor positioning systems based on VLC technology will be able to provide services wherever LEDs are deployed at the price of a minimal overhead.

Light positioning systems will not only benefit the related industry but also a great mass of consumers. Therefore, extensive research has been done within this field. The different positioning algorithms proposed have inherent advantages and drawbacks, and therefore tradeoffs can be made for different applications. We have provided a detailed comparison table based on three metrics: accuracy, spatial dimension, and complexity.

We have also identified challenges for the commercialization of light positioning systems, which include but are not limited to: multipath reflection, synchronization, channel multi-access, service outage, and privacy. Possible solutions are provided herein.

Acknowledgment

The authors sincerely thank the National Science Foundation (NSF) for its great contribution toward the writing of this chapter; NSF Award no. IIP-1169024, IUCRC on optical wireless applications.

References

[1] Federal Communications Commission, "Location-based services: An overview of opportunities and other considerations," [available online]: http://transition.fcc.gov/Daily_Releases/ Daily_Business/2012/db0530/DOC-314283A1.pdf

[2] Foursquare, "About Foursquare," https://foursquare.com/about, 2013.

[3] Pyramid Research, "Navigation providers try to find their way," [available online]: http://www.pyramidresearch.com/points/print/110624.htm

[4] C. Wang, C. Huang, Y. Chen, and L. Zheng, "An implementation of positioning system in indoor environment based on active RFID," in IEEE Joint Conf. on *Pervasive Computing*, pp. 71–76, 2009.

[5] J. Zhou, K. M.-K. Chu, and J. K.-Y. Ng, "Providing location services within a radio cellular network using ellipse propagation model," in IEEE 19th Int. Conf. *Advanced Information Networking and Applications*, pp. 559–564, 2005.

[6] Y. Liu and Y. Wang, "A novel positioning method for WLAN based on propagation modeling," in IEEE Int. Conf. *Progress in Informatics and Computing*, pp. 397–401, 2010.

[7] L. Son and P. Orten, "Enhancing accuracy performance of Bluetooth positioning," in IEEE *Wireless Communications and Networking* Conf., pp. 2726–2731, 2007.

[8] H. Schweinzer and G. Kaniak, "Ultrasonic device localization and its potential for wireless sensor network security," *Control Engineering Practice*, **18**, (*8*), 852–862, 2010.

[9] Y. U. Lee and M. Kavehrad, "Two hybrid positioning system design techniques with lighting LEDs and ad-hoc wireless network," *IEEE Trans. Consum. Electron.*, **58**, (*4*), 1176–1184, 2012.

[10] "Cisco visual networking index: Global mobile data traffic forecast update, 2009–2014," [available online]: http://www.cisco.com/en/US/solutions/collateral/ns341/ns525/ns537/ ns705/ns827/white_paper_c11-520862.html

[11] M. Kavehrad, "Sustainable energy-efficient wireless applications using light," *IEEE Communications Magazine*, **48**, (*12*), 66–73, 2010.

[12] M. Kavehrad, "Optical wireless applications: A solution to ease the wireless airwaves spectrum crunch," in SPIE OPTO Int. Society for Optics and Photonics, pp. 86450G– 86450G, 2013.

[13] H. Liu, H. Darabi, P. Banerjee, and J. Liu, "Survey of wireless indoor positioning techniques and systems," *IEEE Trans. Syst., Man, Cybern. C, Appl. Rev.*, **37**, (*6*), pp. 1067–1080, 2007.

[14] H. Hong, Y. Ren, and C. Wang, "Information illuminating system for healthcare institution," in the 2nd IEEE Int. Conf. on *Bioinformatics and Biomedical Eng.*, pp. 801–804, 2008.

[15] Y. Y. Tan, S. J. Sang, and W. Y. Chung, "Real time biomedical signal transmission of mixed ECG signal and patient information using visible light communication," in 35th Ann. Int. Conf. of the IEEE Engineering in Medicine and Biology Society, pp. 4791–4794, 2013.

[16] T.Q. Wang, Y.A. Sekercioglu, A. Neild, and J. Armstrong, "Position accuracy of time-of-arrival based ranging using visible light with application in indoor localization systems," *J. Lightw. Technol.*, **31**, (*20*), 3302–3308, 2013.

[17] Z. Zhou, M. Kavehrad, and P. Deng, "Indoor positioning algorithm using light-emitting diode visible light communications," *J. of Opt. Eng.*, **51**, (*8*), 085009-1–085009-6, 2012.

[18] Y. Kim, J. Hwang, J. Lee, and M. Yoo, "Position estimation algorithm based on tracking of received light intensity for indoor visible light communication systems," in IEEE 3rd Int. Conf. on *Ubiquitous and Future Networks*, pp. 131–134, 2011.

[19] W. Zhang, M.I.S. Chowdhury, and M. Kavehrad, "An asynchronous indoor positioning system based on visible light communications," *J. of Opt. Eng.*, **53**, (*4*), 045105-1–045105-9, 2014.

[20] A. Küpper, *Location-Based Services: Fundamentals and Operation*, John Wiley and Sons, 2005.

[21] A. Kushki, K.N. Plataniotis, and A.N. Venetsanopoulos, *WLAN Positioning Systems: Principles and Applications in Location-Based Services*, Cambridge University Press, 2012.

[22] K. Panta and J. Armstrong, "Indoor localisation using white LEDs," *Electron. Lett.*, **48**, (*4*), 228–230, Feb. 2012.

[23] S. Jung, S. Hann, and C. Park, "TDOA-based optical wireless indoor localization using LED ceiling lamps," *IEEE Trans. Consum. Electron.*, **57**, (*4*), 1592–1597, 2011.

[24] http://en.wikipedia.org/wiki/Photogrammetry

[25] T. Tanaka and S. Haruyama, "New position detection method using image sensor and visible light LEDs," in IEEE 2nd Int. Conf. on *Machine Vision*, pp. 150–153, 2009.

[26] G.B. Prince and T.D.C. Little, "A two phase hybrid RSS/AoA algorithm for indoor device localization using visible light," in IEEE *Global Commun.* Conf., pp. 3347–3352, 2012.

[27] S.Y. Jung, S. Hann, S. Park, and C.S. Park, "Optical wireless indoor positioning system using light emitting diode ceiling lights," *Microwave and Optical Technology Letters*, **54**, (*7*), 1622–1626, 2012.

[28] G. Yun and M. Kavehrad, "Spot diffusing and fly-eye receivers for indoor infrared wireless communications," in Proc. of IEEE Int. Conf. on *Selected Topics in Wireless Commun.*, pp. 262–265, 1992.

[29] http://en.wikipedia.org/wiki/Global_Positioning_System

[30] W. Elmenreich, "Sensor fusion in time-triggered systems," PhD thesis, Vienna University of Technology, 2002.

5 Visible light positioning and communication

Zhengyuan Xu, Chen Gong, and Bo Bai

5.1 Introduction

Positioning with high accuracy and reliable real-time performance is an urgent need and has become one of the most exciting features of the next generation wireless systems [1–4]. The global positioning system (GPS) suffers from poor performance in certain areas, for example, in indoor environments, due to the strong absorption of the carrier wave by building materials [5], and in urban environments due to link blockage or multipath interference by tall buildings. In such application scenarios, a visible light positioning system can help mobile users to obtain their real-time position information, based on the signal sources from local light infrastructures not being affected by the complicated communication environment.

5.1.1 Indoor light positioning system

In indoor environments, assisted by indoor positioning systems, various location-based services can be realized, such as navigation and guidance in shopping malls, supermarkets, museums, and hospitals; tracking and monitoring of valuable equipments in factories; and improvement of wireless personal network performance through network service adaption. Moreover, multiple optional techniques could be adopted to implement indoor positioning. One straightforward solution is to utilize existing WiFi access points for positioning or to install extra points to ensure wide and dense coverage [6, 7]. This area is still under active research. Another approach is to rely on a densely distributed indoor lighting infrastructure which can serve as location sensing units, such as conventional fluorescent and incandescent lamps. Recently, highly energy-efficient white LEDs have emerged as appealing replacements. Similarly to light positioning systems (LPS) using fluorescent light [8–10], LPS with white LEDs coupled with an inexpensive imaging sensor such as is embedded in a mobile handset [11, 12] is an emerging technology. It can provide concurrent indoor positioning and illumination. The light signal emitted from a modulated white LED, carrying LED position information, is received by a photodetector through a visible light communication (VLC) channel, and then the position of the photodetector is estimated based on the received signal attributes such as amplitude and angle of arrival (AOA). If a receiver is equipped with an image sensor instead of a conventional photodiode or arrayed photodiodes, spatial rejection of interference light from other light sources helps to increase communication signal to noise ratio (SNR) [13, 14]. Meanwhile, due to a large number of small size pixels, high spatial resolution is achievable, leading to a possibly low cost and highly accurate LPS. Since an LPS operates in the visible light spectrum, it does not create electromagnetic interference with existing RF systems critical for RF restricted environments such as hospitals; it is also immune to any RF interference from WiFi or cellular systems.

An LPS using fluorescent light typically can achieve an accuracy of several meters [11] or less than one meter [15, 16] in a small service area. The position calculation algorithms require the incident light signals' horizontal angle, vertical angle and rotation angle, which are usually difficult to measure in practice. On the contrary, an LPS with white LEDs and a camera could achieve a higher accuracy and much wider coverage area of several meters. An indoor positioning system with white LEDs and an image sensor is proposed in [17], but two specifically placed image sensors are required and images of at least four LEDs are supposed to be captured. These requirements raise the system cost and computation complexity. Received light intensity-based positioning methods are proposed in [18, 19], pre-calculated incident light intensity at a given position and at least three white LEDs without inter-cell interference need to be adopted for the trilateration method. Also the unpredictable light sources may bring unavoidable intensity fluctuation, which would make these two systems less reliable.

5.1.2   Outdoor light positioning system

In outdoor environments, the precise position of a vehicle is urgently needed to provide a reliable navigation service in some downtown and urban locations where the most widely used GPS has a poor performance [20] due to link blockage or multipath. Effective methods to improve driver safety and realize an intelligent transportation system (ITS) are necessary. Vision based navigation methods [21, 22] and LED/monocular camera methods [23] have been proposed to achieve the required range finding. With more and more LED-based traffic lights widely deployed in the real world, traffic light-based location beacon and image processing methods [12, 24–27] have also been proposed to obtain vehicle position information. All these methods need an expensive high speed camera and complex image processing procedure to generate the position of a vehicle. Recently, Roberts *et al.* have proposed a vehicle tail lights based method for a light positioning system [28], but it can only generate the relative position of a vehicle to the one in front of it under some restrictive assumptions.

To cater for more general indoor and outdoor application environments, this chapter presents an indoor LPS and outdoor LPS system. For the indoor LPS, we combine VLC with position estimation methods, and let white LEDs emit their location-relevant signals to be captured by a receiving camera. Then an optimal estimation algorithm is implemented at the receiver which can provide an unbiased estimate of the camera position. For the outdoor LPS system, the light signal emitted from the traffic light, carrying its position information, is received by two photodiodes mounted in the front of a vehicle through a visible light communication (VLC) link. Then the position of the vehicle can be estimated based on the received position information of the traffic light and the time difference of arrival (TDoA) of the light signal to the two photodiodes. Methods for LPS with one traffic light and two traffic lights are reviewed, and the positioning error due to non-coplanar effects and the related coplanar rotation methods for both cases are also discussed. To simplify the analysis, the blocking effect of the surroundings, and background light noise and multipath effect in the VLC link are neglected.

5.2   Indoor light positioning systems based on visible light communication and imaging sensors

There are several feasible indoor positioning solutions, based on third generation (3G) systems, WiFi, and the UWB. However, the multiple signal reflections from surrounding objects cause multipath distortion resulting in uncontrollable errors. These effects may degrade all known solutions for indoor positioning which use electromagnetic waves to carry signals from indoor transmitters to indoor receivers. Here we take another approach that records the positions of the LED images, as described below.

5.2.1   System description

Assume that a camera is used as both an image sensor to capture the image points of the white LEDs on the ceiling as well as a communication receiver to capture the position information of the white LEDs carried by the light communication signal. According to Newton's law of lens imaging [29], namely the relationship of the white LEDs and image points on the image sensor of the camera, the position of the camera could be easily estimated. Cases with given white LED positions in the navigation frame and estimated positions obtained by a VLC link are both covered. Considering VLC channel noise [30], camera noise [13, 31], and the attenuation effect of the signal through propagation, the white LED image point on the image sensor may have a random bias and even the position information of the white LED may be misinterpreted at the receiver [32]. Following the concept of localization under noisy observations [33], the current estimation problem can be modeled as an optimal linear combination of noisy measurements. Performance of the mean square error (MSE) is studied based on the Cramer–Rao lower bound (CRLB) as a benchmark measurement.

In a typical LPS with a white LED and camera, the white LED mounted on the ceiling can emit a light signal carrying its position information, usually a unique identity (ID) from which the white LED's position can be obtained, such as the floor number and its location within a floor. An image sensor-based camera assembled in a portable device like a cell phone and personal digital assistant (PDA) could capture the signal light intensity variation where the white LED's position information in the navigation frame resides. Furthermore, the white LED can also be captured by an image sensor as a light point through a properly configured image lens, increasing the FOV of the camera at the same time. A proper optical filter may also be used to reduce the received ambient light in the background.

5.2.2 LPS with known LED positions

Consider an LPS with a camera and N white LEDs mounted on the ceiling whose positions in the navigation system are known as $S_i = [X_i, Y_i, Z_i]^T$, $i = 1, 2, N$, shown in Fig. 5.1. The white LED image points $S'_i = [x_i, y_i]^T$ are captured on the image sensor with the center of the sensor set as the origin of the relative coordinate system. The focal length of the image lens is fixed as f, the distance from the lens of the camera to the image sensor plane is D, and from the lens to the floor is Z_C. Let us focus on the 2-dimensional position vector $p_C = [X_C, Y_C]^T$ of the camera position represented by the center of the image lens **O**, which is the same as the position of the image sensor center in the navigation frame.

Noiseless measurement

Assume the height Z_C of the camera (defined as the height of the image lens) is given. According to the homothetic triangle theory and Newton's law of lens imaging [29], we can easily obtain the position of the camera $p_{Ci} = [X_{Ci}, Y_{Ci}]^T$ from pairs of white LED S_i and its image S'_i on the image sensor as

$$p_{Ci} = [X_{Ci}, Y_{Ci}]^T = [X_i + \lambda_i x_{mi}, Y_i + \lambda_i y_{mi}]^T, \tag{5.1}$$

where

$$\lambda_i = \frac{Z_i - Z_C}{D}, \tag{5.2}$$

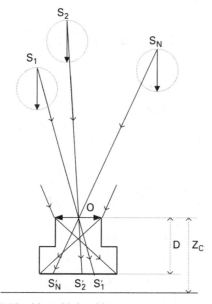

Figure 5.1   LPS with multiple white LEDs and a camera.

and $[x_{mi}, y_{mi}]^T$ is the measured position of the white LED image point $S'_i = [x_i, y_i]^T$ on the image sensor.

Note that, in the absence of noise, N values ($1 \leq i \leq N$) of p_{Ci} obtained from N white LEDs will be identical to $p_C = [X_C, Y_C]^T$.

Noisy measurement

Next we consider the presence of measurement noise for $[x_{mi}, y_{mi}]^T$, where the noise for x_{mi} and y_{mi} satisfies the Gaussian distribution $N(\mu_{xi}, \sigma_{xi}^2)$ and $N(\mu_{yi}, \sigma_{yi}^2)$, respectively, for $1 \leq i \leq N$. Note that this is essentially a parameter estimation problem under various noisy observations, which can be performed via optimal linear combination.

We consider linear combining to estimate $p_C = [X_C, Y_C]^T$ based on the N measurement results p_{Ci} for $1 \leq i \leq N$, given as follows:

$$\hat{X}_C = \frac{\sum_{i=1}^N \beta_{xi}(X_{Ci} - \lambda_i \mu_{xi})}{\sum_{i=1}^N \beta_{xi}}, \tag{5.3}$$

where β_{xi} for $1 \leq i \leq N$ are linear combining coefficients. First, note that $\hat{X}_C$ is an unbiased estimator of X_C, since

$$E\hat{X}_C = \frac{\sum_{i=1}^N \beta_i X_C}{\sum_{i=1}^N \beta_i} = X_C. \tag{5.4}$$

Then, it is desired to find the optimal linear combination coefficients β_{xi}, $1 \leq i \leq N$, that minimize the estimation variance. According to (5.3), the estimation variance is given as follows:

$$\sigma_{XC}^2 = \frac{\sum_{i=1}^N \beta_{xi}^2 \lambda_i^2 \sigma_{xi}^2}{\left(\sum_{i=1}^N \beta_{xi}\right)^2}. \tag{5.5}$$

The following manipulation aims to find the optimal values of $\{\beta_{xi}\}_{i=1}^N$.

Note that

$$\sigma_{XC}^2 = \frac{\sum_{i=1}^N (\beta_{xi}\lambda_i\sigma_{xi})^2}{\left(\sum_{i=1}^N \beta_{xi}\lambda_i\sigma_{xi}\frac{1}{\lambda_i\sigma_{xi}}\right)^2}. \tag{5.6}$$

According to the Cauchy–Schwarz inequality, we have that

$$\left(\sum_{i=1}^N \beta_{xi}\lambda_i\sigma_{xi}\frac{1}{\lambda_i\sigma_{xi}}\right)^2 \leq \sum_{i=1}^N (\beta_{xi}\lambda_i\sigma_{xi})^2 \sum_{i=1}^N \left(\frac{1}{\lambda_i\sigma_{xi}}\right)^2, \tag{5.7}$$

where the equality is achieved if the following is satisfied,

$$\beta_{xi}\lambda_i\sigma_{xi} = \frac{1}{\lambda_i\sigma_{xi}}, \quad \text{for } 1 \leq i \leq N. \tag{5.8}$$

Then, an optimal solution to the linear combination coefficients β_{xi}, denoted as β_{xi}^*, and the minimum variance are given

$$\beta_{xi}^* = \frac{1}{\lambda_i^2 \sigma_{xi}^2}, \ 1 \leq i \leq N, \ \text{and} \ \sigma_{XC}^2 = \frac{1}{\sum_{i=1}^{N} \frac{1}{\lambda_i^2 \sigma_{xi}^2}}. \tag{5.9}$$

Similarly, Y_C can be estimated via computing the following unbiased estimator based on linear combination,

$$\hat{Y}_C = \frac{\sum_{i=1}^{N} \beta_{yi}(Y_{Ci} - \lambda_i \mu_{yi})}{\sum_{i=1}^{N} \beta_{yi}}, \tag{5.10}$$

with the optimal combination coefficients and the minimum variance given

$$\beta_{yi}^* = \frac{1}{\lambda_i^2 \sigma_{yi}^2}, \ 1 \leq i \leq N, \ \text{and} \ \sigma_{YC}^2 = \frac{1}{\sum_{i=1}^{N} \frac{1}{\lambda_i^2 \sigma_{yi}^2}}. \tag{5.11}$$

Cramer–Rao lower bound (CRLB)

CRLB provides a lower bound on the variance of any unbiased estimator [34], and thus can serve as a benchmark for the performance of an unbiased estimator. An unbiased estimator is optimal if it can achieve the CRLB. In the following we show that the above optimal linear combination indeed achieves the CRLB.

Since CRLB provides a lower bound on the variance of an estimator for a deterministic parameter, an unbiased estimator which achieves this lower bound is efficient. More specifically, we have the following joint distribution:

$$p(X_m, X_C) = \prod_{i=1}^{N} \frac{\exp\left[-\frac{(X_{mi} - X_C - \lambda_i \mu_{xi})^2}{2(\lambda_i \mu_{xi})^2}\right]}{\sqrt{2\pi(\lambda_i \sigma_{xi})^2}}. \tag{5.12}$$

We have

$$\frac{\partial^2 \ln p(X_m, X_C)}{\partial X_C^2} = -\sum_{i=1}^{N} \frac{1}{\lambda_i^2 \sigma_{xi}^2}; \tag{5.13}$$

and thus the CRLB for X_C is given by

$$CRLB(X_C) = \frac{1}{\sum_{i=1}^{N} \frac{1}{\lambda_i^2 \sigma_{xi}^2}}. \tag{5.14}$$

Similarly, we have the following CRLB for Y_C,

$$CRLB(Y_C) = \frac{1}{\sum_{i=1}^{N} \frac{1}{\lambda_i^2 \sigma_{yi}^2}}. \tag{5.15}$$

It is seen that the variance of the linear combination can achieve the CRLB, and thus is efficient.

A simple example of the CRLB

We provide a simple example of the CRLB of the position estimation. Assume equal estimation variance for all LEDs in both dimensions x and y, i.e., $\sigma_{xi}^2 = \sigma_{yi}^2 = \sigma^2$ for all $1 \leq i \leq N$; and $\lambda_i = \lambda$ for all $1 \leq i \leq N$. Then the CRLBs for the positions X_C and Y_C can be simplified as follows,

$$CRLB(X_C) = CRLB(Y_C) = \frac{\lambda^2 \sigma^2}{N}. \tag{5.16}$$

The results can be justified by the fact that the average over N measurements reduces the estimation variance by a factor of N.

5.2.3   Monte-Carlo simulation results

Assume four LEDs with positions $(1,1)$, $(-1,-1)$, $(1,-1)$, and $(-1,1)$, and one camera with position $(0,0)$. Assume the height of the four LEDs to the camera $Z_i - Z_C = 3$ for all $1 \leq i \leq 4$; and the distance between the lens of the camera and the image sensor plane $D = 0.1$. Assume zero mean Gaussian measurement noise with variance $\sigma_{xi}^2 = \sigma_{yi}^2 = \frac{\sigma^2}{2}$ for all four LEDs. Figure 5.2 plots the measurement variance averaged over $1 000 000$ random realizations versus the CRLB for $0.005 \leq \sigma \leq 0.05$. It is seen that the measurement variance matches perfectly with the CRLB, which validates the theoretical analysis.

5.3   Outdoor light positioning systems based on LED traffic lights and photodiodes

5.3.1   Light positioning system

We consider an LPS system consisting of one or multiple traffic lights and two photodiodes [35]. Generalization to incorporate more than two photodiodes is straightforward and thus omitted. The light signal emitted from the traffic light, carrying the light position information, is received by two photodiodes mounted in the front of a vehicle. With the TDoA Δt of the light signal to the two photodiodes, the path difference from the traffic light to the two photodiodes is given as $\Delta s = v_L \Delta t$, where v_L is the speed of light. Since a hyperbola is the set of points that have the same distance difference to two fixed points, the traffic light is on a hyperbola determined by the path differences and separation of the

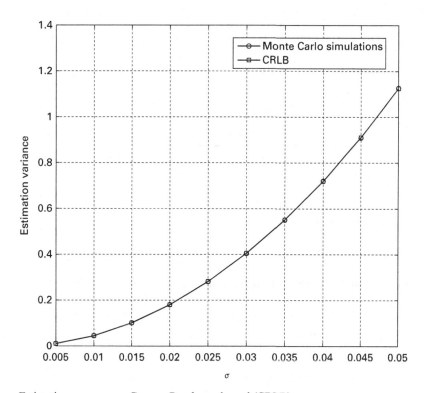

Figure 5.2   Estimation error versus Cramer–Rao lower bound (CRLB).

two photodiodes which are the two foci of the hyperbola [15]. Considering the center of the two photodiodes as the required position of the vehicle, with two hyperbolas determined by one traffic light and two photodiodes in two different locations, or by two traffic lights and two photodiodes, we can derive the relative position of the traffic light to the vehicle, and finally the absolute position of the vehicle. These two cases are discussed below. Here we assume perfect synchronization of the traffic lights that transmit the positioning information, and perfect synchronization of the photo-receivers that receive such information.

LPS with one traffic light

As shown in Fig. 5.3, when the light signal of only one traffic light T_1 is captured by the two photodiodes, we can generate the first hyperbola determined by the obtained TDoA Δt_1 and the distance of the two photodiodes F_1, F_2 in front of a vehicle at time t_1. While the vehicle moves towards the traffic light, the second hyperbola is obtained through TDoA Δt_2 and the same distance of the two photodiodes F'_1, F'_2 at time t_2. The relative position of the traffic light to the vehicle can be derived by the crossing points T_1, T_2, T_3, T_4 of the two hyperbolas and some related constraints given below.

Note that there are different representations of a hyperbola. One familiar representation is that for any point on the hyperbola the distance difference to the two foci is fixed,

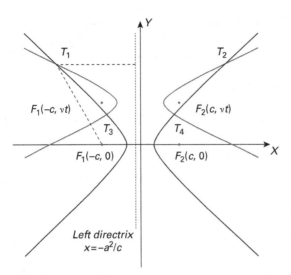

Figure 5.3   LPS with one traffic light at the upper right corner and vehicle movement along the Y-direction.

given by $\|(x,y) - (c,0)\| - \|(x,y) - (-c,0)\| = 2a$. The second form is given mathematically as follows:

$$\frac{x^2}{a^2} - \frac{y^2}{b^2} = 1, \qquad (5.17)$$

where $c^2 = a^2 + b^2$. Parameter a determines the cross-point of the hyperbola to the x-axis. However, in the following, we adopt another form based on a fixed ratio of the distance to one focus over the distance to one line, to obtain a low complexity estimator.

A hyperbola can also be defined as the locus of points whose distance from the focus is proportional to the horizontal distance from a vertical line known as the conic section directrix, and the ratio is the eccentricity, where c is one half of the two photodiodes' separation and $a = \frac{\Delta s}{2}$ is one half of the path difference from the traffic light to the two photodiodes. Let $\|x\| = \sqrt{x^T x}$ denote the square root of norm of vector x. We can obtain the following equations:

$$\|(x,y) - (-c,0)\| = e_1[x - (-\frac{a_1}{e_1})], \qquad (5.18)$$

$$\|(x,y) - (-c,\Delta y)\| = e_2[x - (-\frac{a_2}{e_2})], \qquad (5.19)$$

where (x,y) is the position of the traffic light in the coordinate system whose origin is the center of the two photodiodes, $\Delta y = v_V(t_2 - t_1)$ is the distance the vehicle moves during the time duration and v_V is the speed of the vehicle, $e_i = c/a_i$ $(i = 1,2)$ is the eccentricity. As shown in Fig. 5.3, four crossing points T_1, T_2, T_3, and T_4 are obtained through (5.18) and (5.19). To uniquely specify the desired traffic light T_1, we utilize some prior information. Supposing the two photodiodes are front-oriented without an omnidirectional field of view (FOV), the traffic light is located in front of the vehicle, that is $y > \Delta y$. This thus eliminates the possibility of T_3 and T_4. Furthermore, let us define the TDoA as the time of arrival (ToA) of the light signal from the traffic light to the right photodiode F_2 less the ToA to the left F_1, that is $= t_{T,F2} - t_{T,F1}$. The x-coordinate of the traffic light position is negative ($x < 0$) when the TDoA is positive and $x > 0$ when negative. Supposing the obtained TDoA is positive, the crossing point T_1 is determined as the unique traffic light position. Finally, the absolute position of the vehicle is determined as $(X_0 - x, Y_0 - y)$, where (X_0, Y_0) is the absolute position of the traffic light, which can be obtained by VLC from the traffic light to the photodiodes.

LPS with two traffic lights

In this arrangement, as shown in Fig. 5.4, the light signals of two traffic lights T_1, T_2, for example one traffic light for vehicles and the other for pedestrians, are captured by the two photodiodes F_1, F_2. We can generate two hyperbolas determined by the obtained TDoAs Δt_1 and Δt_2 and the photodiode spacing. With the properties of the hyperbola and the absolute distance and direction of the two traffic lights, we can obtain the following equations:

$$\|(x_1, y_1) - (-c, 0)\| = e_1[x_1 - (-\frac{a_1}{e_1})], \tag{5.20}$$

$$\|(x_2, y_2) - (-c, 0)\| = e_2[x_2 - (-\frac{a_2}{e_2})], \tag{5.21}$$

$$\|(x_1, y_1) - (x_2, y_2)\| = \|(X_1, Y_1) - (X_2, Y_2)\|, \tag{5.22}$$

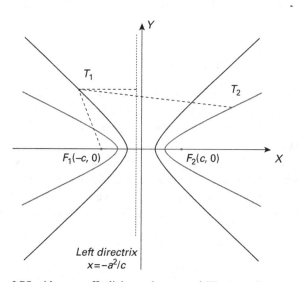

Figure 5.4   LPS with two traffic lights and measured TDoAs at the same time.

$$\frac{x_1 - x_2}{X_1 - X_2} = \frac{y_1 - y_2}{Y_1 - Y_2}, \tag{5.23}$$

where (x_1, y_1) and (x_2, y_2) are the positions of the two traffic lights in the coordinate system with the center of the two photodiodes as the origin, and (X_1, Y_1) and (X_2, Y_2) are the absolute positions of the two traffic lights, respectively. Similarly to the case with one traffic light, some undesirable solutions can be excluded. Note the constraints of $y_1 > 0$, $y_2 > 0$; and $x_1 < 0, x_2 < 0$ when the TDoAs Δt_1 and Δt_2 are positive, and $x_1 > 0, x_2 > 0$ when they are negative. Finally the only possible solution to the absolute position of the vehicle becomes $(X_1 - x_1, Y_1 - y_1)$ or $(X_2 - x_2, Y_2 - y_2)$.

Note that for the case of two traffic lights, one shot of the two TDoAs is taken to estimate the location of the vehicle, while for the case of one traffic light, two shots are necessary. This is due to the fact that the TDoA only specifies a hyperbola on which the location of the lights may lie, and at least two hyperbolas are needed to define a point.

A numerical example on the hyperbola

Assume that two photodiodes are mounted on the front of a vehicle with a 2 meter separation. Consider a coordinate system where the origin lies in the middle of the two photodiodes, with the x-axis perpendicular to the moving direction of the vehicle, and the y-axis aligned with the moving direction of the vehicle. In such a coordinate system, the positions of the two photodiodes are $(-1, 0)$ and $(1, 0)$. Let (x, y) be the position of a light source. Let $\Delta t = 4.10^{-9}$s be the TDoA of the light signal arriving at the two photodiodes, where the arrival time at $(-1, 0)$ is earlier than that at $(1, 0)$ and such that $\|(x, y) - (-1, 0)\| - \|(x, y) - (1, 0)\| = -1.2$ m. It implies that (x, y) lies on a hyperbola with the two foci given by $(-1, 0)$ and $(1, 0)$, which can be expressed in another form by (5.18). More specifically, we have that $c = 1, a_1 = 0.6, b_1 = 0.8$. It is seen that the location of (x, y) lies on the left half of the hyperbola

$$\frac{x^2}{0.36} - \frac{y^2}{0.64} = 1. \tag{5.24}$$

After traveling a distance of 4 m, we have the distances which lead to the distance difference $\|(x, y) - (-1, 4)\| - \|(x, y) - (1, 4)\| = -2$ m. Then we have $c = 1, a_2 = 1, b_2 = 0$, which is an extreme case of a hyperbola. Then, we have $y = 4$, and $x = -0.6\sqrt{26} = -3.06$.

The case of two traffic lights is an extension of that of one light. The location of the vehicle can be determined by the intersection of the two hyperbolas determined by two TDoAs of the signals sent by two lights at one shot. This is essentially of the same nature as the example of localization using one light via several shots, which is given above.

5.3.2   Calibration of error induced by non-coplanar geometry

The two hyperbolas for both LPS with one and two traffic lights, generated in Section 5.3.1, are not necessarily coplanar. This would bring some inevitable positioning error and aggravate the performance of the LPS system, especially when the vehicle is near the traffic light, which would bring a bigger angle between the two hyperbola planes. In this section, we discuss the coplanar rotation method, which is to rotate one plane to coincide with the other around the intersecting line of the two planes, and generate the actual position of the vehicle with the 3-dimensional transform theory.

Coplanar rotation for LPS with one traffic light

For LPS with one traffic light, referring to Fig. 5.5, set the relative position of the traffic light T_1 as (x, y, z) in the coordinate system $XYZO$. The tilt angles of planes determined by the traffic light T_1 and photodiode positions A_1, A_2 at time t_1, and by the traffic light T_1 and photodiode positions B_1, B_2 at time t_2, are $\alpha_1 = \tan^{-1} \frac{z}{y}$ and $\alpha_2 = \tan^{-1} \frac{z}{y - \Delta y}$, respectively. As shown in Fig. 5.5, rotating the plane $T_1 B_1 B_2$ to coincide with the plane $T_1 A_1 A_2$ around the intersecting line MN, we can get the corresponding points B'_1 and B'_2 in plane $T_1 A_1 A_2$ with respect to B_1 and B_2. According to the 3-dimensional transform theory, the positions of the corresponding points B'_1 and B'_2 in the coordinate system $XYZO$ are given by

$$(x'_{B1}, y'_{B1}, z'_{B1}, 1) = (-c, \Delta y, 0, 1) \cdot T \cdot R_x(\Delta\alpha) \cdot T^{-1}, \tag{5.25}$$

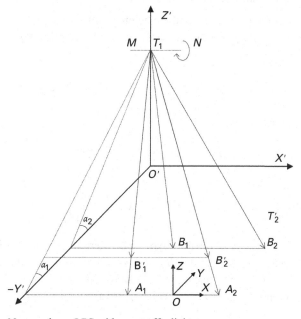

Figure 5.5   Non-coplanar LPS with one traffic light.

$$(x'_{B2}, y'_{B2}, z'_{B2}, 1) = (c, \Delta y, 0, 1) \cdot T \cdot R_x(\Delta \alpha) \cdot T^{-1}, \tag{5.26}$$

where

$$T = \begin{bmatrix} 1 & 0 & 0 & 0 \\ 0 & 1 & 0 & 0 \\ 0 & 0 & 1 & 0 \\ 0 & -y & -z & 1 \end{bmatrix}, \text{ and } R_x(\Delta \alpha) = \begin{bmatrix} 1 & 0 & 0 & 0 \\ 0 & \cos\Delta\alpha & \sin\Delta\alpha & 0 \\ 0 & -\sin\Delta\alpha & \cos\Delta\alpha & 0 \\ 0 & 0 & 0 & 1 \end{bmatrix}, \tag{5.27}$$

are the translation matrix and rotation matrix, respectively, T^{-1} is the inverse matrix of T and $\Delta\alpha = \alpha_2 - \alpha_1$ is the rotation angle. Note that here we only rotate the coordinate on the y and z-axes clockwise, such that the rotation via angle $\Delta\alpha$ is added on the y and z coordinates. The rotation angle can be found from α_1 and α_2. We have that

$$\sin\alpha_1 = \frac{z}{\sqrt{z^2 + y^2}}, \quad \cos\alpha_1 = \frac{y}{\sqrt{z^2 + y^2}},$$
$$\sin\alpha_2 = \frac{z}{\sqrt{z^2 + (y - \Delta y)^2}}, \quad \cos\alpha_2 = \frac{y - \Delta y}{\sqrt{z^2 + (y - \Delta y)^2}}, \tag{5.28}$$

from which we can derive the values of $\sin\Delta\alpha$ and $\cos\Delta\alpha$. More specifically, we have that

$$\cos\Delta\alpha = \cos(\alpha_2 - \alpha_1) = \cos\alpha_2 \cos\alpha_1 + \sin\alpha_2 \sin\alpha_1,$$
$$\sin\Delta\alpha = \sin(\alpha_2 - \alpha_1) = \sin\alpha_2 \cos\alpha_1 - \cos\alpha_2 \sin\alpha_1, \tag{5.29}$$

based on results, $R_x(\Delta\alpha)$, obtained.

When the height H of the traffic light and the height h of the photodiodes are known, we can obtain $z = H - h$. And the TDoA of the mapped traffic light T'_2 to the two photodiodes is

$$a'_2 = \frac{1}{2}(\|(x, y, z) - (x'_{B1}, y'_{B1}, z'_{B1})\| - \|(x, y, z) - (x'_{B2}, y'_{B2}, z'_{B2})\|). \tag{5.30}$$

Then, (5.18) and (5.19) can be written as

$$\|(x, y, z) - (0, 0, -c)\| = e_1 \left[x - \left(-\frac{a_1}{e_1} \right) \right], \tag{5.31}$$

$$\|(x, y, z) - (x'_{B1}, y'_{B1}, z'_{B1})\| = e'_2 \left[x - \left(-\frac{a'_2}{e'_2} \right) \right], \tag{5.32}$$

where $e'_2 = \frac{c}{a'_2}$ is the eccentricity of the new hyperbola determined by the traffic light T_1 and the two mapped photodiode positions B'_1 and B'_2. Combined with the constraints given in Section 5.3.1, we can determine the actual relative position of the traffic light and then the absolute position of the vehicle as $(X_0 - x, Y_0 - y)$.

Coplanar rotation for LPS with two traffic lights

Similarly to the LPS with one traffic light, for the LPS with two traffic lights shown in Fig. 5.6, rotating the plane $T_2A_1A_2$ to coincide with the plane $T_1A_1A_2$ around the intersecting line OX, we can get the corresponding points T'_2 in plane $T_1A_1A_2$ with respect to T_2. The relative positions of the traffic lights T_1 and T_2 are (x_1, y_1, h_1) and (x_2, y_2, h_2) in the coordinate system $XYZO$, and absolute positions are (X_1, Y_1, H_1) and (X_2, Y_2, H_2), respectively. According to the 3-dimensional transform theory, the relative and absolute positions of the corresponding point T'_2 are given by

$$(x'_2, y'_2, h'_2, 1) = (x_2, y_2, h_2, 1) \cdot R_x(\Delta\alpha), \tag{5.33}$$

$$(X'_2, Y'_2, H'_2) = (X_2, Y_2, H_2) - (x_2 - x'_2, y_2 - y'_2, h_2 - h'_2), \tag{5.34}$$

and the TDoA of the mapped traffic light T'_2 to the two photodiodes A_1, A_2 is

$$a'_2 = \frac{1}{2}\left[\sqrt{(x'_2 + c)^2 + (y'_2)^2 + (h'_2)^2} - \sqrt{(x'_2 - c)^2 + (y'_2)^2 + (h'_2)^2} \right]. \tag{5.35}$$

Then, (5.20)–(5.23) can be rewritten as

$$\|(x_1, y_1, h_1) - (-c, 0, 0)\| = e_1\left[x_1 - \left(-\frac{a_1}{e_1}\right) \right], \tag{5.36}$$

$$\|(x'_2, y'_2, h'_2) - (-c, 0, 0)\| = e'_2\left[x'_2 - \left(-\frac{a'_2}{e'_2}\right) \right], \tag{5.37}$$

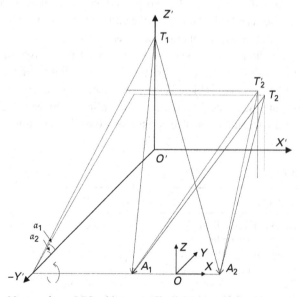

Figure 5.6   Non-coplanar LPS with two traffic lights.

$$\|(x_1, y_1, h_1) - (x'_2, y'_2, h'_2)\| = \|(X_1, Y_1, H_1) - (X'_2, Y'_2, H'_2)\|, \qquad (5.38)$$

$$\frac{x_1 - x'_2}{X_1 - X'_2} = \frac{y_1 - y'_2}{Y_1 - Y'_2} = \frac{h_1 - h'_2}{H_1 - H'_2}, \qquad (5.39)$$

where $e'_2 = \frac{c}{d'_2}$ is the eccentricity of the new hyperbola determined by the mapped traffic light T'_2 and the two photodiodes. Combined with the constraints given in Section 5.3.1, we can generate the actual relative position of the traffic light and then the absolute position of the vehicle as $(X_1 - x_1, Y_1 - y_1)$ or $(X'_2 - x'_2, Y'_2 - y'_2)$.

5.3.3   Numerical results

In this section, we evaluate the performance of the above LPS methods, both with and without coplanar rotation. We adopt the following estimation bias, namely the mean positioning error as performance metric:

$$Bias = \sqrt{Bias_X^2 + Bias_Y^2}, \qquad (5.40)$$

where $Bias_X$ and $Bias_Y$ are the LPS bias errors in the X-axis and Y-axis, respectively. Assuming that the light signal propagates in an ideal noise-free VLC link and the light signal is well synchronized, there would be no error introduced in the absolute position information obtained on the traffic light source and the TDoA of the light signal from the traffic light to the two photodiodes. The traffic light T_1 for vehicles and the traffic light T_2 for pedestrians on a crossing are located 3 m left and 1 m right of the center of a vehicle in the X-axis, respectively. The two photodiodes are mounted on the front of the vehicle with a 2 m separation, and the middle point of the two photodiode positions is treated as the position of the vehicle. The vehicle moves towards the traffic light on a flat road and records the TDoA of the light signal to the two photodiodes every 0.1 s. The heights of the traffic lights T_1 and T_2 and photodiodes and some other basic parameters are given in Table 5.1.

Figure 5.7 shows the LPS bias error performance with one traffic light against the distance from the traffic light to the vehicle in the Y-axis. The LPS bias without coplanar rotation grows significantly when the distance from the traffic light to the vehicle in the Y-axis is reduced, especially when the distance is less than 20 m. The positioning bias gets

Table 5.1 Basic parameters.

Symbol	Quantity	Value
H_1	Height of traffic light T_1	6 m
H_2	Height of traffic light T_2	4 m
h	Height of photodiodes	1 m
D	Separation of two photodiodes	2 m
$t_2 - t_1$	Recorded time duration	0.1 s

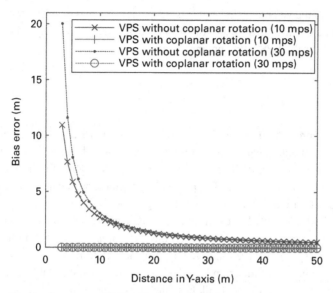

Figure 5.7   Bias performance of LPS with one traffic light.

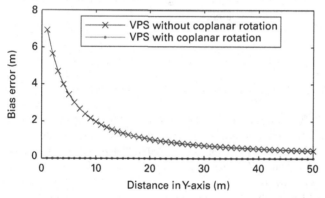

Figure 5.8   Bias performance of LPS with two traffic lights.

bigger when the vehicle speeds up from 10 m/s to 30 m/s, and there are still 0.50 m and 0.52 m positioning errors with 10 m/s and 30 m/s even when the distance is up to 50 meters. But there is no positioning bias with coplanar rotation as no error is introduced in the measurement as assumed.

Now suppose the two traffic lights T_1 and T_2 are the same distance from the vehicle in the Y-axis. Figure 5.8 shows the bias performance of the LPS with two traffic lights against the distance from the traffic light to the vehicle in the Y-axis. The positioning bias of the LPS without coplanar rotation increases dramatically when the distance is reduced. The bias error, reduced to 0.4 m for the distance of 50 m, grows more slowly than the LPS with one traffic light while the distance decreases. It becomes 3.4 m when the distance of the traffic light to the vehicle on the Y-axis is 5 m, where the bias error is 5.9 m and 8.1 m for LPS with one traffic light at speeds of 10 m/s and 30 m/s, respectively. Also there is no bias introduced for the LPS with coplanar rotation.

5.4 Summary

This chapter first presents an indoor LPS model using white LEDs and a camera, and the corresponding optimal linear combination for an unbiased estimate of the camera position that achieves the CRLB. Then it presents a traffic light and photodiode based LPS model for vehicle positioning. The position of a vehicle can be determined based on the received position information of the traffic light and the TDoA of the light signal to the two photodiodes mounted in the front of the vehicle. Methods for LPS with one and two traffic lights are both presented. When the non-coplanar effect is introduced, the vehicle positioning methods are improved by coplanar rotation.

This chapter considers indoor and outdoor positioning using visible light. A basic assumption here is the line-of-sight link from the visible light source, without any optical wireless communication link error. Positioning based on optical wireless multipath received signals, especially in the scenario of indoor positioning with wall reflection or outdoor positioning with joint surrounding object reflection and vehicle movement, remains a subject for future research.

References

[1] Gu, Y., Lo, A. & Niemegeers, I. (2009), "A survey of indoor positioning systems for wireless personal networks," *IEEE Trans. Communications Surveys and Tutorials* **11**, 13–32.

[2] Liu, H., Darabi, H., Banerjee, P. & Liu, J. (2007), "Survey of wireless indoor positioning techniques and systems," *IEEE Trans. Systems, Man, and Cybernetics, Part C: Applications and Reviews* **37**, 1067–1080.

[3] Cheok, A. & Li, Y. (2008), "Ubiquitous interaction with positioning and navigation using a novel light sensor-based information transmission system," *Personal and Ubiquitous Computing* **12**, 445–458.

[4] Cheok, A. & Li, Y. (2010), "A novel light-sensor-based information transmission system for indoor positioning and navigation," *IEEE Trans. Instrumentation and Measurement* **60**, 290–299.

[5] Khoury, H. M. & Kamat, V. R. (2009), "Evaluation of position tracking technologies for user localization in indoor construction environments," *Automation in Construction* **18**, 444–457.

[6] Woo, S., Jeong, S., Mok, E. *et al.* (2011), "Application of WiFi-based indoor positioning system for labor tracking at construction sites: A case study in Guangzhou MTR," *Automation in Construction* **20**, 3–13.

[7] Lashkari, A. H., Parhizkar, B. & Ngan, M. N. A. (2010), "WiFi-based indoor positioning system," in 2nd International Conference on *Computer and Network Technology*, Bangkok, pp. 76–78.

[8] Liu, X., Makino, H., Kobayashi, S. & Maeda, Y. (2006), "An indoor guidance system for the blind using fluorescent lights – relationship between receiving signal and walking speed," in Proc. 28th *Engineering in Medicine and Biology* Society Conference, New York, pp. 5960–5963.

[9] Liu, X., Umino, E. & Makino, H. (2009), "Basic study on robot control in an intelligent indoor environment using visible light communication," in 6th IEEE International Symposium on *Intelligent Signal Processing*, Budapest, pp. 417–428.

[10] Randall, J., Amft, O., Bohn, J. & Burri, M. (2007), "Luxtrace: Indoor positioning using building illumination," *Personal and Ubiquitous Computing* 11, 417–428.

[11] Yoshino, M., Haruyama, S. & Nakagawa, M. (2008), "High-accuracy positioning system using visible LED lights and image sensor," in IEEE *Radio and Wireless* Symposium, Orlando, pp. 439–442.

[12] Randall, J., Amft, O., Bohn, J. & Burri, M. (2003), "Positioning beacon system using digital camera and LEDs," *IEEE Trans. Vehicular Technology* 52, 406–419.

[13] Bigas, M., Cabruja, E., Forest, J. & Salvi, J. (2006), "Review of CMOS image sensors," *Microelectronics Journal* 37, 433–451.

[14] Castillo-Vazquez, M. & Puerta-Notario, A. (2005), "Single-channel imaging receiver for optical wireless communications," *IEEE Communications Letters* 9, 897–899.

[15] Liu, X., Makino, H., Kobayashi, S. & Maeda, Y. (2008), "Research of practical indoor guidance platform using fluorescent light communication," *IEICE Transactions on Communications* E91B, 3507C3515.

[16] Sertthin, C., Ohtsuki, T. & Nakagawa, M. (2010), "6-axis sensor assisted low complexity high accuracy-visible light communication based indoor positioning system," *IEICE Transactions on Communications* E93B, 2879–2891.

[17] Shaifur, R. M., Haque, M. M. & Kim, K. (2011), "High-accuracy positioning system using visible LED lights and image sensor," in 14th International Conference on *Computer and Information Technology (ICCIT)*, Dhaka, pp. 309–314.

[18] Kim, Y., Hwang, J., Lee, J. *et al.* (2011), "Position estimation algorithm based on tracking of received light intensity for indoor visible light communication systems," in 3rd International Conference on *Ubiquitous and Future Networks (ICUFN)*, pp. 131–134.

[19] Kim, H., Kim, D., Yang, S., Son, Y. & Han, S. (2011), "Indoor positioning system based on carrier allocation visible light communication," in *Lasers and Electro-Optics*, Pacific Rim, Sydney, pp. 787–789.

[20] Savasta, S., Pini, M. & Marfia, G. (2008), "Performance assessment of a commercial GPS receiver for networking applications," in 5th IEEE *Consumer Communications and Networking* Conference, pp. 613–617.

[21] Savasta, S., Joo, T. & Cho, S. (2006), "Detection of traffic lights for vision-based car navigation system," in 1st Pacific Rim Symposium, PSIVT, pp. 682–691.

[22] Wang, W. & Cui, B. (2006), "Automatic monitoring and measuring vehicles by using image analysis," in Proceedings of *SPIE*-IS and T *Electronic Imaging*, pp. 1–8.

[23] Watada, S., Hayashi, K., Toda, M. *et al.* (2009), "Range finding system using monocular in-vehicle camera and LED," in *Intelligent Signal Processing and Communication Systems*, 2009, IEEE, pp. 493–496.

[24] Nagura, T., Yamazato, T., Katayama, M. *et al.* (2010*a*), "Improved decoding methods of visible light communication system using LED array and high-speed camera," in 71st *Vehicular Technology* Conference (VTC 2010-Spring), pp. 1–5.

[25] Nagura, T., Yamazato, T., Katayama, M. *et al.* (2010*b*), "Tracking an led array transmitter for visible light communications in the driving situation," in 7th International Symposium on *Wireless Communication Systems* (*ISWCS*), pp. 765–769.

[26] Pang, G. & Liu, H. (2001), "Led location beacon system based on processing of digital images," *IEEE Trans. Intelligent Transportation Systems* **2**, 135–150.

[27] Pang, G., Liu, H., Chan, C. & Kwan, T. (1998), "Vehicle location and navigation systems based on leds," in Proceedings of 5th World Congress on *Intelligent Transport Systems*, Seoul, pp. 12–16.

[28] Roberts, R., Gopalakrishnan, P. & Rathi, S. (2010), "Visible light positioning: Automotive use case," in IEEE *Vehicular Networking* Conference (VNC), pp. 309–314.

[29] Hecht, E. (2001), *Optics* (4th ed.), Addison Wesley.

[30] Cui, K., Chen, G., Xu, Z. & Roberts, R. D. (2010), "Line-of-sight visible light communication system design and demonstration," in Proc. of 7th IEEE, IET International Symposium on *Communication Systems, Networks and Digital Signal Processing*, Newcastle, pp. 21–23.

[31] Gow, R., Renshaw, D., Findlater, K. *et al.* (2007), "A comprehensive tool for modeling CMOS image-sensor-noise performance," *IEEE Trans. Electron. Devices* **54**, 1321–1329.

[32] Sadler, B. M., Liu, N., Xu, Z. & Kozick, R. (2008), "Range-based geolocation in fading environments," in Proc. of *Allerton* Conference, Monticello, pp. 23–26.

[33] Liu, N., Xu, Z. & Sadler, B.M. (2008), "Low complexity hyperbolic source localization with a linear sensor array," *IEEE Signal Processing Letters* **15**, 865–868.

[34] Poor, H. V. (1994), *An introduction to signal detection and estimation*, Springer.

[35] Bai, B., Chen, G., Xu, Z. & Fan, Y. (2011), "Visible light positioning based on LED traffic light and photodiode," in IEEE *Vehicular Technology* Conference (VTC Fall), pp. 1–5.

6 The standard for visible light communication

Kang Tae-Gyu

6.1 Scope of VLC standard

Visible light communication VLC has the advantage that the emitting source of wireless communication uses LED light [1]. The standard for illumination has been developed covering connections between lamp and electronic power in terms of electric safety, as set out in IEC TC 34. The standard for visible light communication needs a number of protocols between the sending party and the receiving party, as given in PLASA E1.45 [2] and IEEE 802.15.7 [3]; as well as dealing with electric safety. We have to consider compatibilities even though the VLC service area, illumination, vendor, and standard are different.

6.1.1 VLC service area compatibility

Visible light communication services can be provided in different illumination space areas [12]. The area could be a lighting space for a museum, shopping mall, hallway, office, restaurant, etc. There are two VLC service styles. One is for a specific area, such as a company or organisation defined location in which proprietary equipment can be used. The other is a public area where in order to communicate the equipment must be compatible with communications standards. When we design for a specific area, we do not need any standard for VLC. Design of a specific VLC can be done easily without any constraints or limitations because the design is based on proprietory technology and not on a defined standard such as (4) in Fig. 6.1. This specific style VLC has the advantage of fast, cheap deployment at the first opening stage, but has the disadvantage of lacking VLC service area compatibility such as (2), (5), (6) in Fig. 6.1.

We need standards to ensure VLC service area compatibility for any kind of service area such as (1) and (3) in Fig. 6.1. The international standards are IEEE 802.15.7 [7] and PLASA E1.45 [11]. The domestic or national standards consist of three documents in Japan [8] and 18 documents in Korea [9].

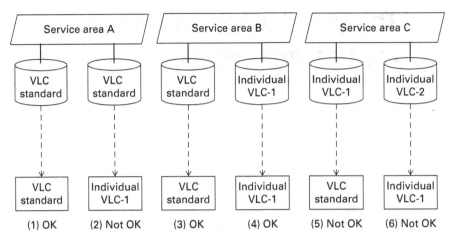

Figure 6.1   Example of VLC service area compatibility.

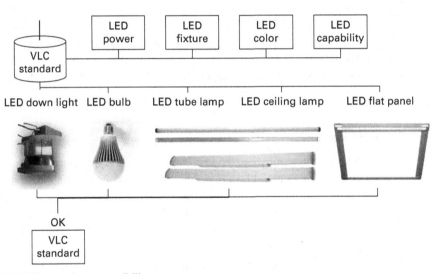

Figure 6.2   VLC illumination compatibility.

6.1.2   VLC illumination compatibility

There are various LED illumination power capabilities, fixtures, and colors; their capabilities and their shapes vary according to their intended usage. Although there is a wide range of LED illumination systems, the visible light communication standard has to be applicable to all kinds of LED system.

We need to know the LED illumination standards in IEC TC 34, so as to go on to develop new standards, including those for VLC components in LED illumination, see Fig. 6.2.

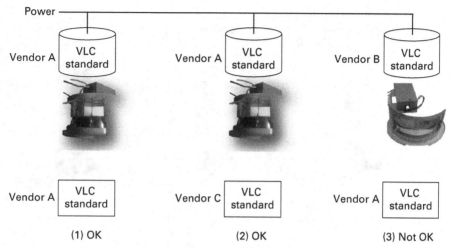

Figure 6.3   VLC vendor compatibility.

6.1.3   VLC vendor compatibility

There are many LED illumination vendors, who can introduce and withdraw LED illumination products freely. Illumination and receiving terminals may cease to function or reach the end of their operational life span. The VLC standard has to support VLC vendor compatibility. We can choose LED illumination and receiving terminals transparently without selecting specific vendors or products. We can replace at any time, any product, any manufacturer, and any vendor such as (1) and (2) in Fig. 6.3. In case (3) in the figure, there is a problem due to lack of vendor compatibility.

In order to achieve VLC vendor compatibility, we need a standard with inter-operable profiles. If we cannot buy a suitable product with VLC compatibility, we have to choose re-installation of illumination and replace the receiving terminal or abandon the VLC service and install only illumination lighting. We cannot abandon the lighting function, which is mandatory in our lives.

6.1.4   Standard compatibility

There are several standards relating to visible light communication: IEEE 802.15.7 VLC PHY/MAC; IEC TC 34 LED illumination; PLASA E1.45 DMX-512A VLC; and LED lighting source Zhaga engine [10]. IEEE 802.15.7 VLC PHY/MAC, issued in 2011, covers VLC PHY interior LED illumination and VLC PHY receivers, see Fig. 6.4.

The IEC TC 34 LED illumination intelligent system lighting (ISL) ad hoc Working Group is developing standards for digital functional components including visible light communication. PLASA E1.45 DMX-512A VLC issued in 2013 covers visible light communication data, and wired transfer between LED illumination and a control server. Zhaga engine applies to LED lighting sources for illumination. LED lighting

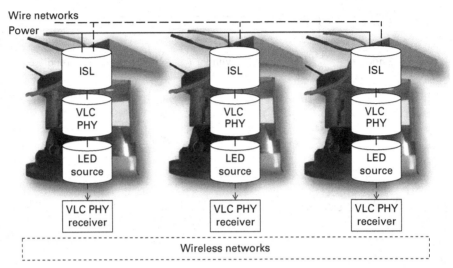

Figure 6.4   Visible light communication standard compatibility.

on and off switching can be controlled using wireless networks: ZigBee, IrDA, Bluetooth, and WiFi.

International or domestic standard specifications can be developed concurrently by many standard organizations or working groups, who can share their standards' activities and draft specifications via exchanging documents. This should ensure standard compatibility. A new standard specification might upgrade or "version up" a previous one, and must be written to ensure forward and backward standard compatibility. The result is consistent principles for developers and end users and avoidance of confusion.

6.2   VLC modulation standard

IEEE 802.15.7 VLC has three different PHY (physical layers) that depend on the application: PHY I, PHY II, and PHY III. PHY I is intended for outdoor use with low data rate applications [4–6]. This mode uses on-off keying (OOK) and variable pulse position modulation (VPPM) with data rates in the tens to hundreds of kb/s. PHY II is intended for indoor use with moderate data rate applications. This mode uses OOK and VPPM with data rates in the tens of Mb/s. PHY III is intended for applications using color shift keying (CSK) that have multiple light sources and detectors with data rates in the tens of Mb/s.

6.2.1   Variable pulse position modulation VPPM

VLC needs a modulation strategy that is compatible with dimming control of illumination. One example of such a strategy is variable pulse position modulation (VPPM).

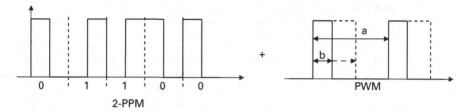

Figure 6.5   Principle of variable pulse position modulation VPPM.

VPPM is a modulation scheme that is compatible with dimming control that varies the duty cycle or pulse width to achieve dimming, as opposed to amplitude control.

VPPM combines 2-PPM with PWM for a dimming control. Bits "1" and "0" in VPPM are distinguished by the position of a pulse, whereas the width of the pulse is determined by the dimming ratio. The principles of VPPM are illustrated in Fig. 6.5.

6.2.2   Line coding

The 4B6B line code expands each block of four bits into an encoded block of six bits with DC balance. This means there will always be precisely three zeros and three ones in each block of six encoded bits.

6.3   VLC data transmission standard

There are two types of visible light communication data transmission. One is fixed data, and the other is changeable data within lighting for visible light communication. The changeable data can be changed in accordance with wired transmission protocols and wireless transmission protocols.

6.3.1   Wired transmission protocol

There are two kinds of candidate wired visible light communication data transmission protocol: PLASA E1.45 DMX-512A VLC; and IEC 62386 DALI VLC (Table 6.1). PLASA is a trade association for Professional Lighting and Sound with a technical standards program.

PLASA E1.45 DMX-512A VLC was issued in 2013 to allow communication of 802 data to luminaires over an ANSI E1.11 DMX512-A data link for data transmission from those luminaires using visible light communication, IEEE 802.15.7. ANSI E1.11, revised in 2008, describes a method of digital data transmission for control of lighting equipment and accessories, including dimmers, color-changers, and related equipment. DMX512-A can be used for outdoor media façade LED lighting [13].

Table 6.1 Wired transmission protocol for visible light communication.

Standard specification	Organization	Functions
E1.45 DMX-512A VLC	PLASA	DMX512-A data link for data transmission from those luminaires using visible light communication, IEEE 802.15.7
IEC 62386 DALI	IEC TC 34	A protocol for control by digital signals of electronic lighting equipment

IEC TC 34 62386 DALI is one of the candidate protocols for visible light communication transmission data. The International Electro-technical Commission (IEC) is the world organization that prepares and publishes international standards for all electrical, electronic and related technologies [12]. IEC 62386 digital addressable lighting interface (DALI) specifies a protocol for control by digital signals of electronic lighting equipment. DALI can be used for indoor lighting dimming control.

6.3.2   Wireless transmission protocol

There are complementary protocols for wireless transmission: ZigBee, IrDA, Bluetooth, and wireless LAN. Visible light communication needs additional wireless communication skills due to its unidirectional communication principles.

ZigBee, defined in IEEE 802.15.4, is used in applications that require only a low data rate, long battery life, and secure networking with 250 kbit/s transfer rate. ZigBee can be used for wireless light switches including dimming controls.

6.4   VLC illumination standard

The advantage of visible light communication technology is its usage of LED illumination without any other transmission media, so we need to apply traditional illumination standards. TC 34 prepares international standards for lamps and other related equipment, and was established in 1948.

6.4.1   LED lighting source interface

The international Zhaga Consortium is developing interface specifications that enable interchangeability of LED light sources made by different manufacturers. The Zhaga specifications known as books, describe the interfaces between LED luminaires and LED light engines. This will accelerate the adoption of LED lighting solutions in the marketplace.

Eight books of specifications cover the technology as follows: 1 overview, 2, 5, 6, 8 socketable drum-shaped LED sources, 3 circular LED modules, 4 and 7 rectangular LED modules according to shape of luminaire.

Visible light communication uses LED light, but the Zhaga LED module has not yet been considered as visible light communication. When we develop a visible light communication PHY and an application service, we have to consider Zhaga LED module specifications.

6.4.2   Fixture interface

International Electro-technical Commission Technical Committee TC 34 has been developing international standards regarding specifications for lamps including LEDs, lamp caps and holders, lamp control gear, luminaires, and miscellaneous related equipment not covered by projects of other technical committees.

Visible light communication can be one of a luminaire's functions. IEC TC 34 has not yet developed any specifications for visible light communication. They are needed to specify how to merge lighting itself and wireless communication using light.

6.4.3   LED intelligent system lighting interface

The IEC TC 34 intelligent system lighting ad hoc Working Group had a first face-to-face meeting in January 2014. Its remit is new creative convergence technology between the traditional lighting industry and information and communications technology (ICT). The function of LED lighting has introduced use of ICTs such as wireless communication, wired communication, and visible light communications, see Fig. 6.6. Wireless communication technologies can be adapted using ZigBee, Bluetooth, IrDA, and wireless LAN according to application specific requirements.

Visible light communication requires development of co-operative luminaire functions and other information technology such as wireless communication and wired communication.

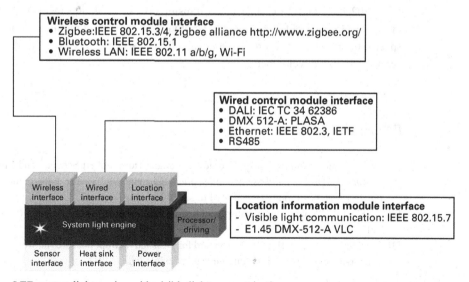

Figure 6.6   LED system light engine with visible light communication.

6.4.4   VLC service standard

We can find service standard activities for visible light communication in IEEE 802.15.7, IEC TC 34, PLASA CPWG, TTA VLC WG, VLCC, and ITU-T SG 16.

The book *Advanced Optical Wireless Communication* Chapter 14 [1] names as visible light communication applications: VLC guidance systems, VLC color imaginable systems, VLC indoor navigator, and VLC automobile driving support systems.

A VLC guidance system uses lamps that illuminate a yard, national border, or facility, for guidance as well as for protection from outside attacks. The lamps have an identification number (VLC ID or LED ID) and guidance information. A VLC color imaginable system uses color lamps for color information itself, whether from instinct or education. A VLC indoor navigator uses lamps with visible light communication for indoor sales area navigation in locations where GPS is not supported. A VLC automobile driving support system uses lamps including headlights, fog lights, turn signals, and brake lights for safe driving.

The VLC Working Group (WG4021) in TTA started in May 2007. VLC WG developed TTA 5 VLC standard specifications in 2008, covering: Basic Configurations of Transmitter PHY for Visible Light Communication, Basic Configurations of Receiver PHY for Visible Light Communication, Basic Configuration of LED Interface for Illumination and Visible Light Communication, Basic Configuration of Light Location Information Service Model using Visible Light Communication, and Basic Configuration of Lighting Identification for Visible Light Communication.

VLC WG developed TTA 23 VLC and LED control related standard drafts in 2013. The 18 specifications are focused on the ways to combine visible light communication and illumination technologies.

Visible light communication can deliver creative services, but relevant standard specifications have not yet been developed. However, the VLC PHY specification in IEEE 802.15.7, and VLC data wired transmission specification in PLASA E1.45 are available. Opening of the visible light communications market requires service standard specifications. These will provide guidance for users from the viewpoint of application service function development.

References

[1] Shlomi Arnon, John R. Barry, George K. Karagiannidis, Robert Schober, and Murat Uysal, *Advanced Optical Wireless Communication*, Chapter 14 "Visible light communication," pp. 351–368, Cambridge University Press, 2012.

[2] ANSI E1.45, "Unidirectional transport of IEEE 802 data frames over ANSI E1.11 (DMX512-A)," 2013.

[3] IEEE Std 802.15.7-2011, "IEEE standard for local and metropolitan area networks – part 15.7: Short-range wireless optical communication visible light," 2011.

[4] Sang-Kyu Lim, "ETRI PHY proposal on VLC band plan and modulation schemes for illumination," IEEE 802.15-09-0674-00-0007, 2009.

[5] Dae Ho Kim, "ETRI PHY proposal on VLC line code for illumination," IEEE 802.15-09-0675-00-0007, 2009.

[6] Youjin Kim, "Analysis of IP-based control networks for LED lighting fixture communication," *New Trends in Information Science and Service Science (NISS)*, 4th IEEE Conference, 2010, pp. 307–312.

[7] Eun Tae Won, (2009, January), IEEE 802.15 WPAN™ Task Group 7 (TG7) Visible Light Communication, Available: http://www.ieee802.org/15/pub/TG7.html.

[8] Masao Nakagawa, (2007), *Visible Light Communication Consortium*, Available: http://www.vlcc.net.

[9] Leem Chasik, (2014), Telecommunication Technology Association Visible Light Convergence Communication Project Group 425, Available: http://www.tta.or.kr/English/.

[10] "LED light sources interchangeable," Zhaga Consortium, (2014), Available: http://www.zhagastandard.org/.

[11] Karl Ruling, (2014), PLASA Standards, Control Protocol Working Group, Available: https://www.plasa.org/.

[12] International Electrotechnical Commission Technical Sub-committee 34C, "Auxiliaries for lamps," (2014), http://www.iec.ch/.

[13] Sang-Kyu Lim, "Entertainment lighting control network standardization to support VLC services," *IEEE Communication Magazine* **51**, (12), pp. 42–48, 2013.

7 Synchronization issues in visible light communication

Shlomi Arnon

In the preceding chapters, many facets of VLC have been outlined demonstrating the promising features of this emerging illumination and communication technology. It is anticipated that VLC will become a familiar facility in the office, on the road (V2V communications) and even in toys (Disney Research). The broad range of VLC applications is driving an exponential rise in the required data rate, while the semiconductor industry is keeping pace in developing light sources that could provide the necessary high data rate. The high data rate VLC paradigm opens up new opportunities for modulation methods that are being developed to meet the particular requirements of simultaneous illumination and communication. Demodulation of the signal and, hence, extraction of the transmitted information require stringent synchronization. In this chapter, we present four VLC modulation methods: on off keying (OOK), pulse position modulation (PPM), inverse pulse position modulation (IPPM), and variable pulse position modulation (VPPM), and develop expressions that describe the bit error rates (BER) for each method. We provide examples for the effect of clock time shift and jitter on the system BER performance for the inverse pulse position modulation (IPPM) method.

7.1 Introduction

The many different applications of VLC systems, such as V2V [1–5], short range communication for toys [6–8] and high data rate indoor implementations [7–13] have been elaborated upon in the preceding chapters. While the underlying VLC principle is simple – an electrical signal from the transmitter is converted to a modulated optical signal by the light source and carried by the illuminating light to the receiver, different modulation methods that encapsulate the information within the illumination light are possible. The common modulation methods can be divided into three families [11, 13, 14], based on a) illumination color e.g. color shift keying (CSK), b) many subcarriers e.g. discrete multitone (DMT) or orthogonal frequency division multiplexing (OFDM), and c) intensity measured in the time domain, including on off keying (OOK), pulse position modulation (PPM), inverse PPM (IPPM) and variable pulse position modulation (VPPM). These methods make it possible to use the illumination light as a tool to carry

Parts of this chapter are based on the article by Shlomi Arnon, "The effect of clock jitter in visible light communication applications," *Journal of Lightwave Technology*, 30, (*21*), 3434–3439, 2012

information in a relatively straightforward way. However, the performance of time domain modulation methods is extremely sensitive to time synchronization and clock jitter. In this chapter, we explain the four different time domain modulation methods mentioned above and develop expressions for the BER that can be achieved. The remainder of this chapter expands on the effect of time synchronization on the performance of one of these methods (IPPM).

7.2   VLC modulation methods in the time domain

In this section we describe in more detail four different modulation methods that encapsulate information in the time domain in the context of VLC. The main difference between the conventional communication scenarios and VLC is the requirement to maintain adequate communication performance alongside the necessity to dim the light intensity. We now review a) OOK, b) PPM, c) IPPM, and d) VPPM.

7.2.1   On off keying (OOK)

On off keying (OOK) is a form of binary signal modulation in which a bit is encoded by transmitting optical power for T seconds if the information is "1," whereas when the information is "0," no optical power is transmitted in the T-second time-slot [13]. Dimming of the light is implemented by reducing the transmitted pulse power in accordance with the required percentage dimming (Fig. 7.1).

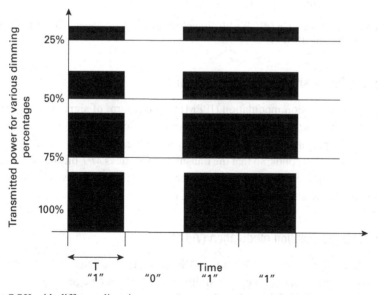

Figure 7.1   OOK with different dimming percentages, where the encoded information is "1 0 1 1."

7.2.2   Pulse position modulation (PPM)

Pulse position modulation (PPM) [13, 14, 15] is a form of signal modulation in which M message bits are encoded by transmitting a pulse for the duration of $T_C = T/2^M$ seconds in one of the 2^M possible time-shifts within a time-slot of T seconds, which is the symbol duration. This scheme is repeated every T seconds, so that the transmitted bit rate is M/T bits per second. Dimming of the light is implemented by reducing the transmitted pulse power in accordance with the required percentage of dimming (Fig. 7.2).

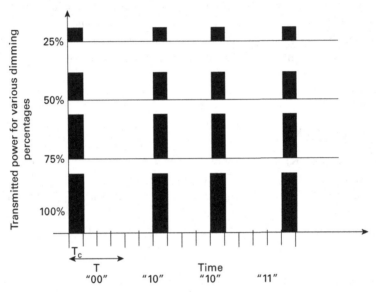

Figure 7.2   Pulse position modulation (PPM) with different dimming percentages, where the encoded information is "00 10 10 11."

7.2.3   Inverse pulse position modulation (IPPM)

Inverse pulse position modulation (IPPM) [16] is a form of signal modulation in which M message bits are encoded by transmitting power for T seconds, except for a "hole" in one of the 2^M possible time-shifts. The duration of the "hole" is $T_C = T/2^M$. This scheme is repeated every T seconds, so that the transmitted bit rate is M/T bits per second. Dimming of the light is implemented by reducing the transmitted pulse power in accordance with the required percentage dimming (Fig. 7.3).

7.2.4   Variable pulse position modulation (VPPM)

Variable pulse position modulation (VPPM) is a form of signal modulation in which the message bit is encoded by transmitting a pulse at the beginning of the symbol for "0" and at the end of the symbol for "1" [11, 14, 17]. The duration of the pulse is

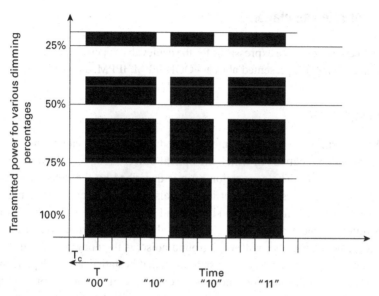

Figure 7.3   Inverse pulse position modulation (IPPM) with different dimming percentages, where the encoded information is "00 10 10 11."

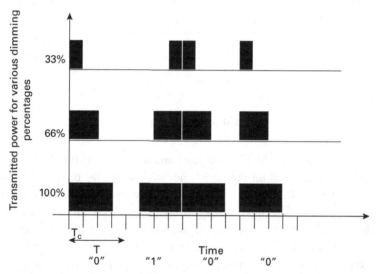

Figure 7.4   Variable pulse position modulation (VPPM) with different dimming percentages, where the encoded information is "0 1 0 0."

determined in accordance with the percentage of required illumination (dimming). The main advantage of this method is that the communication is not affected by a change of the amount of dimming, as long as there is some illumination. This scheme is repeated every T seconds, so that the transmitted bit rate is $1/T$ bits per second (Fig. 7.4).

7.3   Bit error rate calculation

In this section we derive expressions for the achievable bit error rate (BER) with the four modulation methods presented above – OOK, PPM, IPPM, and VPPM.

7.3.1   OOK BER

OOK is a form of binary signal intensity modulation, which can be implemented with direct detection receivers in which the optical power is converted to electronic signals by a photodetector. The conversion ratio is described by the detector's responsivity, R. We denote the resultant electrical signals, prior to decision-making, by y. The receiver integrates the received signals and the decision whether a 1 or a 0 was transmitted is made according to a given algorithm criterion. We assume that the electronic noise, together with background noise, is the dominant noise source and it is modeled by additive white Gaussian noise that is statistically independent between time-slots. The noise has zero mean and covariance of σ_1^2 and σ_0^2 for signals 1 and 0, respectively. After integration, the signals are described by the following conditional densities:

$$P(y|\text{"1"}) = \frac{1}{\sqrt{2\pi}\sigma_1}e^{-\frac{(y-\mu_1)^2}{2\sigma_1^2}} \tag{7.1}$$

and

$$P(y|\text{"0"}) = \frac{1}{\sqrt{2\pi}\sigma_0}e^{-\frac{(y-\mu_0)^2}{2\sigma_0^2}}. \tag{7.2}$$

In the next section, synchronization jitter and offset are analyzed; therefore we now rearrange the expression for the signal and noise so that the bit/symbol duration, T, is part of the signal. The signal is described by the square root of the energy, so that $\mu_1 = \eta_{\text{Dimming}} R P_1 T^{0.5}$ and $\sigma_1^2 = \sigma_{TH}^2 + 2q\eta_{\text{Dimming}} RP_1$ are the received optical signal and the accompanying receiver noise standard deviation, respectively. $\mu_0 = 0$ and $\sigma_0^2 = \sigma_{TH}^2$ are the received optical signal and the receiver noise standard deviation when no power is received, respectively, P_1 is the optical power, σ_{TH}^2 includes the effects of thermal noise and of background shot noise and η_{Dimming} is the illumination dimming factor $0 < \eta_{\text{Dimming}} \leq 1$.

The decision algorithm in this case is based on the maximum a-posteriori probability (MAP) criterion, which maps the received signal according to the following:

$$\hat{s} = \underset{s}{MAX}\left\{\frac{P(y|s)P(s)}{P(y)}\right\}, \tag{7.3}$$

where $P(y|s)$ is the conditional probability that if a bit s is transmitted (taking one of two values, 1 or 0), a y will be received, $P(s)$ is the a-priori probability that a 1 or a 0 is transmitted and $P(y)$ is the a-priori probability of y.

The denominator is identical for all signals and therefore does not affect the decision. In communication systems, the probabilities of transmitting 1 and 0 bits are, in most cases, equal, so we can simply invoke (7.3) and use the maximum likelihood (ML) estimator.

In that case, the likelihood function is given by

$$\Lambda(y) = \frac{P(y/on)}{P(y/off)}$$

$$= \frac{\sigma_{TH}}{\sqrt{\sigma_{TH}^2 + 2q\eta_{\text{Dimming}}RP_1}} \exp \left(\begin{array}{c} -y^2 \left(\frac{1}{2(\sigma_{TH}^2 + 2q\eta_{\text{Dimming}}RP_1)} - \frac{1}{2\sigma_{TH}^2} \right) \\ + \left(\frac{y\eta_{\text{Dimming}}RP_1\sqrt{T}}{(\sigma_{TH}^2 + 2q\eta_{\text{Dimming}}RP_1)} \right) \\ - \frac{(y\eta_{\text{Dimming}}RP_1\sqrt{T})^2}{2(\sigma_{TH}^2 + 2q\eta_{\text{Dimming}}RP_1)} \end{array} \right).$$

$$(7.4)$$

In the case where $\sigma^2_{TH} \gg 2q\eta_{\text{Dimming}}RP_1$, Eq. (7.4) could be simplified by taking the natural logarithm, $\ln(x)$, of both sides of the equation, canceling common factors, and rearranging. Due to the complexity of this expression, it is common to use an approximation for the bit error probability (BER) given by

$$BER \approx \frac{1}{2} erfc \left(\frac{\eta_{\text{Dimming}}RP_1\sqrt{T}}{\sqrt{2}\left(\sqrt{\sigma_{TH}^2 + 2q\eta_{\text{Dimming}}RP_1} + \sqrt{\sigma_{TH}^2} \right)} \right). \quad (7.5)$$

7.3.2   PPM BER

Pulse position modulation (PPM) is a form of signal intensity modulation which can be implemented with direct detection receivers in which the optical power is converted to electronic signals by a photodetector, as for OOK. The resultant electrical signals, before a decision is made, are given by y_i, where $i \in \{0 \ldots 2^M - 1\}$. The receiver integrates the received signals in each of the 2^M possible time-shifts. We assume that the electronic noise, together with background noise, is the dominant noise source when no optical signal is transmitted (0) and the signal shot noise, together with electronic noise and background noise, is the dominant noise source otherwise. In both cases the noise can be modeled by additive white Gaussian noise that is statistically independent between time-slots. The noise has zero mean and a covariance of σ_1^2 and σ_{0i}^2 for 1 and 0, respectively. After integration, the signals y_i, are described by the following conditional densities:

$$P(y_0|\text{"1"}) = \frac{1}{\sqrt{2\pi}\sigma_1} e^{-\frac{(y_0 - \mu_1)^2}{2\sigma_1^2}} \qquad (7.6)$$

and

$$P(y_i|\text{"0"}) = \frac{1}{\sqrt{2\pi}\sigma_{0i}} e^{-\frac{(y_i - \mu_{0i})^2}{2\sigma_{0i}^2}} \quad i \in \{1 \ldots 2^M - 1\}, \qquad (7.7)$$

where $\mu_1 = \eta_{\text{Dimming}} R\, P_1 T_C^{0.5}$ and $\sigma_1^2 = \sigma_{TH}^2 + 2q\eta_{\text{Dimming}} RP_1$ are the received optical signal and the accompanying receiver noise standard deviation, respectively, for a transmitted 1 and $\mu_{0i} = 0$ and $\sigma_{0i}^2 = +\sigma_{TH}^2$ are the received optical signal and the receiver noise standard deviation for a transmitted 0, respectively. R is the responsivity of the detector, P_1 is the optical power, σ_{TH}^2 includes the effects of thermal noise and of background shot noise and η_{Dimming} is the illumination dimming factor $0 < \eta_{\text{Dimming}} \leq 1$.

At the end of the integration period, the receiver makes a decision as to which slot has the largest value by comparing the 2^M integration results. The decision algorithm calculates the indices of the minimum values of the signal vector, which is described mathematically by the function [J,B] = max(A). This function finds the indices of the maximum values of A and returns them in the output vector J, while B is the largest element in A. A is given by $A = [\mu_0 \ldots \mu_{2M-1}]$. A practical implementation of the decision algorithm could employ a comparator that finds the largest voltages from 2^M slot measurements at the end of the symbol period. The calculation of the BER can be done easily if we first calculate the probability of a correct decision. The correct decision is calculated by examining the received signal amplitudes over a range from minus infinity to infinity. For each value in this range we calculate the probability that the received signal in the pulse slot will be larger than the value of the signal from all other time-slots. Hence, 1 minus the calculated correct probability is the erroneous probability. We also assume that all y_i for $i > 1$ are identical and independently distributed (i.i.d) random processes, so they can be represented by an identical density function such as $P(y) = \frac{1}{\sqrt{2\pi}\sigma_{0i}} e^{-\frac{(y-\mu_{0i})^2}{2\sigma_{0i}^2}}$.

In that case the BER is given by

$$BER = \frac{2^{M-1}}{(2^M - 1)} \left[1 - \left[\int_{-\infty}^{\infty} \frac{1}{\sqrt{2\pi}\sigma_1} e^{-\frac{(x-\mu_1)^2}{2\sigma_1^2}} \left(\int_{-\infty}^{x} \frac{1}{\sqrt{2\pi}\sigma_{0i}} e^{-\frac{(y-\mu_{01})^2}{2\sigma_{0i}^2}} dy \right)^{2^M - 1} dx \right] \right].$$

$$(7.8)$$

The term $2^{M-1}/(2^M - 1)$ is used to convert the symbol error rate to the bit error rate. Employing the error function, defined as $erf(x) = \frac{2}{\sqrt{\pi}} \int_0^x \exp(-t^2)dt$, we obtain the following BER expression:

$$BER = \frac{2^{M-1}}{(2^M - 1)} \left[1 - \left[\int_{-\infty}^{\infty} \frac{1}{\sqrt{2\pi}\sigma_1} e^{-\frac{(x-\mu_1)^2}{2\sigma_1^2}} \left(1 + erf\left(\frac{(x-\mu_{0i})}{\sqrt{2}\sigma_{0i}} \right) \right)^{2^M - 1} dx \right] \right].$$

(7.9)

7.3.3   IPPM BER

IPPM is a form of signal intensity modulation, which can be implemented with direct detection receivers in which the optical power is converted to electronic signals by a photodetector, as for OOK and PPM. The resultant electrical signals, before a decision is made, are given by y_i where $i \in \{0 \ldots 2^M - 1\}$. The receiver integrates the received signals in each of the 2^M possible time-shifts. At the end of the integration period the receiver makes a decision as to which slot has the smallest value by comparing the 2^M integration results. We assume that the electronic noise, together with background noise, is the dominant noise source during the "hole" and the "signal" shot noise, together with electronic noise and background noise, is the dominant noise source otherwise. In both cases the noise can be modeled by additive white Gaussian noise that is statistically independent between time-slots. The noise has zero mean and covariance of σ_i^2 and σ_0^2 for a signal and a hole, respectively. After integration, the signals x, and y_i are described by the following conditional densities:

$$P(y_0|\text{``hole''}) = \frac{1}{\sqrt{2\pi}\sigma_0} e^{-\frac{(y_0 - \mu_0)^2}{2\sigma_0^2}}$$

(7.10)

and

$$P(y_i|\text{``signal''}) = \frac{1}{\sqrt{2\pi}\sigma_i} e^{-\frac{(y_i - \mu_1)^2}{2\sigma_i^2}} \quad i \in \{1 \ldots 2^M - 1\},$$

(7.11)

where $\mu_i = \eta_{\text{Dimming}} R P_1 T_C^{0.5}$ and $\sigma_i^2 = 2q\eta_{\text{Dimming}} RP_1 + \sigma_{TH}^2$ are the received optical signal and the accompanying receiver noise standard deviation, respectively, and $\mu_0 = 0$ and $\sigma_0^2 = \sigma_{TH}^2$ are the received optical signal and the receiver noise standard deviation when a hole is received, respectively. R is the responsivity of the detector, P_1 is the optical power, σ_{TH}^2 includes the effects of thermal noise and of background shot noise, and η_{Dimming} is the illumination dimming factor $0 < \eta_{\text{Dimming}} \leq 1$.

The decision algorithm calculates the indices of the minimum values of the signal vector and is given mathematically by the function [J,B] = min(A). This function finds the indices of the minimum values of A and returns them in the output vector J, while B is the smallest element in A. A is given by $A = [\mu_0 \ldots \mu_{2M-1}]$. Practical implementation of the decision algorithm could employ a comparator that finds the lowest voltages from 2^M slot measurements at the end of the symbol period. The calculation of the BER can be done easily if we first calculate the probability of a correct decision. The correct decision is calculated by examining the received signal amplitudes over a range from minus infinity to infinity. For each value in this range we calculate the probability that the received signal in the hole slot will be smaller than the value of the signal from all other time-slots. Hence, 1 minus the calculated correct probability is the erroneous probability. We also assume that all y_i for $i \in \{1 \ldots 2^M-1\}$ are i.i.d random processes so they can be represented by an identical density function such as $P(y) = \frac{1}{\sqrt{2\pi}\sigma_i} e^{-\frac{(y-\mu_i)^2}{2\sigma_i^2}}$.

In that case the BER is given by

$$BER = \frac{2^{M-1}}{(2^M-1)} \left[1 - \left[\int_{-\infty}^{\infty} \frac{1}{\sqrt{2\pi}\sigma_0} e^{-\frac{(x-\mu_0)^2}{2\sigma_0^2}} \left(\int_{x}^{\infty} \frac{1}{\sqrt{2\pi}\sigma_i} e^{-\frac{(y-\mu_1)^2}{2\sigma_1^2}} dy \right)^{2^M-1} dx \right] \right].$$

(7.12)

The term $2^{M-1}/(2^M-1)$ is used to convert the symbol error rate to the BER.

7.3.4   VPPM BER

VPPM is a form of binary signal intensity modulation, which can be seen to be a combination of pulse width modulation (PWM), which is used to control the illumination, and PPM, which is used to carry communication information. As with OOK, PPM, and IPPM, direct detection receivers in which the optical power is converted to electronic signals by a photodetector are used. The resultant electrical signals, before a decision is made, are given by y. The receiver integrates the received signals in the first slot and the last slot of the symbol. At the end of the integration period the receiver decides which slot has the largest value. We assume that the electronic noise, together with background noise, is the dominant noise source when no power is transmitted, and the signal shot noise, together with electronic noise and background noise, is the dominant noise source otherwise. In both cases the noise can be modeled by additive white Gaussian noise that is statistically independent between time-slots. The noise has zero mean and covariance of σ_i^2 and σ_0^2 for signal and hole, respectively. After integration, the signals y, are described by the following conditional densities:

$$P(y|\text{``Left\_slot''}) = \frac{1}{\sqrt{2\pi}\sigma_L} e^{-\frac{(y-\mu_L)^2}{2\sigma_L^2}}$$

(7.13)

and

$$P(y|\text{``Right\_slot''}) = \frac{1}{\sqrt{2\pi}\sigma_R} e^{-\frac{(y-\mu_R)^2}{2\sigma_R^2}}.$$

(7.14)

When 1 is transmitted, $\mu_R = R\,P_L T_C^{0.5}$, $\sigma_R^2 = 2qRP_1 + \sigma_{TH}^2$, $\mu_L = 0$ and $\sigma_L^2 = \sigma_{TH}^2$ are the received optical signal and the accompanying receiver noise standard deviation, respectively, for the right and left slots. When 0 is transmitted $\mu_L = R\,P_L T_C^{0.5}$, $\sigma_L^2 = 2qRP_1 + \sigma_{TH}^2$, $\mu_R = 0$ and $\sigma_R^2 = \sigma_{TH}^2$ are the received optical signal and the accompanying receiver noise standard deviation, respectively, for the left and right slots. The BER is given by

$$
BER = \left[\begin{array}{l} \left(1 - \int_{-\infty}^{\infty} \frac{1}{\sqrt{2\pi}\sigma_R} e^{-\frac{(y_R - \mu_R)^2}{2\sigma_R^2}} \int_{-\infty}^{y_R} \frac{1}{\sqrt{2\pi}\sigma_L} e^{-\frac{(y_L - \mu_L)^2}{2\sigma_L^2}}\, dy_L \right) dy_R \Bigg|_{"1"} + \\[2em] \left(1 - \int_{-\infty}^{\infty} \frac{1}{\sqrt{2\pi}\sigma_L} e^{-\frac{(y_L - \mu_L)^2}{2\sigma_L^2}} \int_{-\infty}^{y_L} \frac{1}{\sqrt{2\pi}\sigma_R} e^{-\frac{(y_R - \mu_R)^2}{2\sigma_R^2}}\, dy_R \right) dy_L \Bigg|_{"0"} \end{array} \right].
$$

$$(7.15)$$

The above equation can be simplified, yielding

$$
BER = \left[1 - \left[\int_{-\infty}^{\infty} \frac{1}{\sqrt{2\pi(\sigma_{TH}^2 + 2qRP_1)}} e^{-\frac{(x - RP_1\sqrt{T})^2}{2(\sigma_{TH}^2 + 2qRP_1)}} \left(1 + erf\left(\frac{x}{\sqrt{2\sigma_{TH}^2}} \right) \right) dx \right] \right].
$$

$$(7.16)$$

7.4   The effect of synchronization time offset on IPPM BER

We now examine in depth the effect of clock synchronization in one of the four modulation schemes introduced in the preceding sections. In order to define the required performance of the clock we derive a mathematical model that describes the effect of clock jitter on the performance of the BER model in (7.12). We follow similar assumptions to those outlined in [16, 18, 19] in order to derive the mathematical model of the BER with clock jitter. The primary assumption in (7.12) is that the time-slot timing is perfect and the decoder in the switch integrates the signals exactly over the T-second interval that constitutes a symbol. If a time offset of Δ seconds occurs during a symbol period, due to timing errors in synchronization, then the integration occurs over an offset time interval. That is, the decoder starts and stops integration over a T-second interval that is displaced by Δ seconds from that containing the actual symbol information. Assuming, without loss of generality, a positive timing offset ($0 < \Delta < T_C$) and equiprobable time-slots, the various effects on the integration statistics are summarized in Table 7.1 and Fig. 7.5. As a result of the time displacement, only a portion of the true signal energy is included in the signal integration, while some signal energy contributes to the integration in the adjacent time-slot, causing inter-symbol interference (ISI) in the form of energy

Table 7.1 Statistical parameters of the IPPM method with time offset. (Courtesy of IEEE, reprinted with permission from [16].)

N	Subsequent symbol "hole" is located in first time-slot	Transmitted symbol location of the "hole" (u)	Slot	Conditional densities' parameters μ	σ^2	Probability
I	No	$0<u<2^M - 1$	"hole"	$\mu_{I1}=R\,P_1\Delta^{0.5}$	$\sigma^2_{I1}=2qRP_1\Delta/T_C+\sigma^2_{TH}$	$((2^M - 1)/2^M)^2$
			Adjacent slot to "hole"	$\mu_{I2}=R\,P_1(T_C-\Delta)^{0.5}$	$\sigma^2_{I2}=2qRP_1(T_C-\Delta)/T_C+\sigma^2_{TH}$	
			Non-adjacent slot to "hole"	$\mu_{I3}=R\,P_1T_C^{0.5}$	$\sigma^2_{I3}=2qRP_1+\sigma^2_{TH}$	
II	No	$u=2^M$	"hole"	$\mu_{II1}=R\,P_1\Delta^{0.5}$	$\sigma^2_{II1}=2qRP_1\Delta/T_C+\sigma^2_{TH}$	$(2^M - 1)/(2^M)^2$
			Adjacent slot to "hole"	$\mu_{II2}=R\,P_1(T_C-\Delta)^{0.5}$	$\sigma^2_{II2}=2\,qRP_1(T_C-\Delta)/T_C+\sigma^2_{TH}$	
			Non-adjacent slot to "hole"	$\mu_{II3}=R\,P_1T_C^{0.5}$	$\sigma^2_{II3}=2qRP_1+\sigma^2_{TH}$	
III	Yes	$0<u<2^M - 1$	"hole"	$\mu_{III1}=R\,P_1\Delta^{0.5}$	$\sigma^2_{III1}=2qRP_1\Delta/T_C+\sigma^2_{TH}$	$(2^M - 1)/(2^M)^2$
			Adjacent slot to "hole"	$\mu_{III2}=R\,P_1\,(T_C-\Delta)^{0.5}$	$\sigma^2_{III2}=2qRP_1(T_C-\Delta)/T_C+\sigma^2_{TH}$	
			Non-adjacent slot to "hole"	$\mu_{III3}=R\,P_1T_C^{0.5}$	$\sigma^2_{III3}=2qRP_1+\sigma^2_{TH}$	
			Slot at 2^M location	$\mu_{III4}=R\,P_1\,(T_C-\Delta)^{0.5}$	$\sigma^2_{III4}=2qRP_1(T_C-\Delta)/T_C+\sigma^2_{TH}$	
IV	Yes	$u=2^M$	"hole"	$\mu_{IV1}=0$	$\sigma^2_{IV1}=\sigma^2_{TH}$	$1/(2^M)^2$
			Adjacent slot to "hole"	$\mu_{IV2}=R\,P_1(T_C-\Delta)^{0.5}$	$\sigma^2_{IV2}=2qRP_1(T_C-\Delta)/T_C+\sigma^2_{TH}$	
			Non-adjacent slot to "hole"	$\mu_{IV3}=R\,P_1T_C^{0.5}$	$\sigma^2_{IV3}=2qRP_1+\sigma^2_{TH}$	

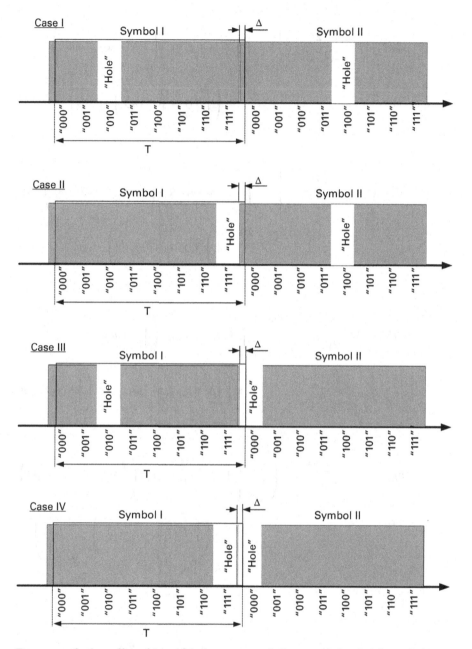

Figure 7.5   Four cases of a time offset of Δ seconds that can occur during a symbol period due to timing errors in synchronization (courtesy of IEEE [16]).

spillover. The effect of this interference depends upon the form of the adjacent time-slot, i.e. whether it contains signal energy or not. The bit error probability for a positive timing error, Δ, for each of the four cases is given in the four equations (7.17)–(7.20). The parameters for each equation are given in Table 7.1.

$$BER_I(\Delta) = \frac{2^M}{2(2^M - 1)} \left[1 - \left[\int\limits_{-\infty}^{\infty} N(x, \mu_{I1}, \sigma_{I1}) \left(\int\limits_{x}^{\infty} N(y, \mu_{I2}, \sigma_{I2}) dy \right) \right. \right.$$

$$\left. \left. \times \left(\int\limits_{x}^{\infty} N(y, \mu_{I3}, \sigma_{I3}) dy \right)^{2^M - 2} dx \right] \right],$$

(7.17)

$$BER_{II}(\Delta) = \frac{2^M}{2(2^M - 1)} \left[1 - \left[\int\limits_{-\infty}^{\infty} N(x, \mu_{II1}, \sigma_{I1}) \left(\int\limits_{x}^{\infty} N(y, \mu_{II2}, \sigma_{II2}) dy \right) \right. \right.$$

$$\left. \left. \times \left(\int\limits_{x}^{\infty} N(y, \mu_{II3}, \sigma_{II3}) dy \right)^{2^M - 2} dx \right] \right],$$

(7.18)

$$BER_{III}(\Delta) = \frac{2^M}{2(2^M - 1)} \left(1 - \left[\int\limits_{-\infty}^{\infty} N(x, \mu_{III1}, \sigma_{III1}) \left(\int\limits_{x}^{\infty} N(y, \mu_{III2}, \sigma_{III2}) dy \right) \right. \right.$$

$$\left. \left. \times \left(\int\limits_{x}^{\infty} N(y, \mu_{III3}, \sigma_{III3}) dy \right)^{2^M - 3} \left(\int\limits_{x}^{\infty} N(y, \mu_{III4}, \sigma_{III4}) dy \right) dx \right] \right),$$

(7.19)

$$BER_{IV}(\Delta) = \frac{2^M}{2(2^M - 1)} \left(1 - \left[\int\limits_{-\infty}^{\infty} N(x, \mu_{IV1}, \sigma_{IV1}) \left(\int\limits_{x}^{\infty} N(y, \mu_{IV2}, \sigma_{IV2}) dy \right) \right. \right.$$

$$\left. \left. \times \left(\int\limits_{x}^{\infty} N(y, \mu_{IV3}, \sigma_{IV3}) dy \right)^{2^M - 2} dx \right] \right),$$

(7.20)

where

$$N(z, \mu, \sigma) = \frac{1}{\sqrt{2\pi}\sigma} e^{-\frac{(z - \mu)^2}{2\sigma^2}}.$$

(7.21)

Averaging over all the possibilities given in Table 7.1, the average BER is given by

$$BER_{avg}(\varepsilon) = \frac{(2^M - 1)^2}{2^{2M}} BER_I(\varepsilon) + \frac{2^M - 1}{2^{2M}} BER_{II}(\varepsilon) + \frac{2^M - 1}{2^{2M}} BER_{III}(\varepsilon)$$

$$+ \frac{1}{2^{2M}} BER_{IV}(\varepsilon).$$

(7.22)

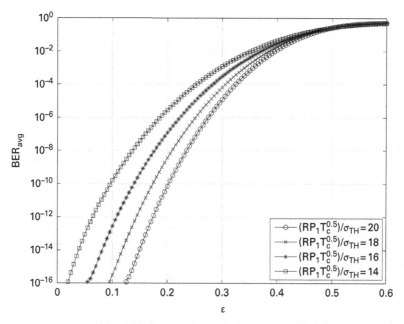

Figure 7.6   Average BER as a function of ε (courtesy of IEEE [16]).

Figure 7.6 depicts the average BER as a function of ε, where ε is the percentage timing error and is given by $\varepsilon = \Delta/T_C$. The IPPM order is $2^M = 16$ with $M = 4$. The results show a relatively fast increase in BER (system performance degradation) as the offset ε is increased. It is easy to see that, defining a BER equal to 10^{-6} as the acceptable limit, the system performance is essentially ruined when ε is increased above 0.19, 0.22, 0.25, and 0.275 for $RP_1 T_C^{0.5}/\sigma_{TH}$ values of 14, 16, 18, and 20, respectively.

7.4.1   The effect of clock jitter on IPPM BER

As we have seen in the previous section, it is crucial to synchronize the encoder to the beginning of the symbol, as otherwise the system performance will severely deteriorate. The common method of synchronizing communication systems is based on synchronization circuits, such as a phased locked loop (PLL). A PLL includes a voltage control oscillator (VCO) and a phase detector. However, as system noise is part of any practical physical system, clock jitter is unavoidable. Therefore, the error probability for the synchronized system is the expectation of Eq. (7.22) with respect to the distribution of clock jitter. We assume that the clock jitter can be modeled by a symmetrical distribution around zero; in that case, the expectation of BER_{avg} is given by:

$$E[BER_{avg}] = 2\int_0^\infty BER_{avg}(\Delta)f(\varepsilon)d\varepsilon, \qquad (7.23)$$

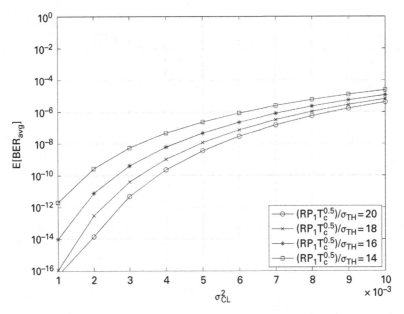

Figure 7.7 $E[BER_{avg}]$ versus the clock jitter variance (courtesy of IEEE [16]).

where $f(x)$ is the probability density function of the clock jitter. The clock timing error jitter is assumed to have a Gaussian distribution with zero mean and variance σ^2_{CL}:

$$f(\varepsilon) = \frac{1}{\sqrt{2\pi}\sigma_{CL}} e^{-\frac{\varepsilon^2}{2\sigma^2_{CL}}}. \tag{7.24}$$

Equation (7.23) was evaluated numerically and the resulting $E[BER_{avg}]$ is plotted in Fig. 7.7 versus the clock jitter variance. The results show a severe degradation of receiver performance with increasing timing error variance. It is easy to see that $E[BER_{avg}]$ increases almost exponentially for a linear increase in σ^2_{CL}. It is important to emphasize that $E[BER_{avg}]$ deteriorates for very small values of clock jitter variance due to the fact that $E[BER_{avg}]$ is not a linear function.

Analyzing the simulation results under the assumption that $M = 4$ and information data rates are 38.4 Mbps [11], 96 Mbps [11] and 1 Gbps results in values of $T = 26.04$ nsec, $T = 10.04$ nsec and $T = 1$ nsec, as well as $T_C = 6.5$ nsec, $T_C = 2.6$ nsec and $T_C = 250$ psec, respectively. In that case, for $RP_1 T_C^{0.5}/\sigma_{TH} = 20$ and $BER = 10^{-6}$, the σ_{CL} should be better than 582 psec, 232 psec and 22 psec, respectively, for the given data rate, i.e. nearly 2 orders of magnitude lower than the symbol period, T.

7.5   Summary

In this chapter we have described modulation methods that can be used in VLC and expanded on the issue of synchronization. The IPPM solution was examined and it was demonstrated that proper clock function is critical in order to provide accurate synchronization and to achieve the required BER.

The main objective in VLC design is to achieve enhanced lighting efficiency. However, in many cases the efficiency decreases when the switching rate is increased. In addition, the time response of phosphor-based lighting systems is quite slow. As a result, the light source modulation speed is low and has a large, finite response time. These factors limit the communication speed and contribute to inter-symbol interference (ISI). With current technology, the modulation method format discussed does not offer high data rate, but it is anticipated that a new generation of LEDs could address this limitation and be modulated at very high speeds – up to 10 GHz [20, 21].

References

[1] Shlomi Arnon, "Optimised optical wireless car-to-traffic-light communication," *Transactions on Emerging Telecommunications Technologies* **25**, 660–665, 2014.

[2] Seok Ju Lee, Jae Kyun Kwon, Sung-Yoon Jung, and Young-Hoon Kwon, "Evaluation of visible light communication channel delay profiles for automotive applications," *EURASIP Journal on Wireless Communications and Networking* (*1*), 1–8, 2012.

[3] Sang-Yub Lee, Jae-Kyu Lee, Duck-Keun Park, and Sang-Hyun Park, "Development of automotive multimedia system using visible light communications," in *Multimedia and Ubiquitous Engineering*, Springer, pp. 219–225, 2014.

[4] S.-H. Yu, Oliver Shih, H.-M. Tsai, and R. D. Roberts, "Smart automotive lighting for vehicle safety," *Communications Magazine*, IEEE **51**, (*12*), 50–59, 2013.

[5] Shun-Hsiang You, Shih-Hao Chang, Hao-Min Lin, and Hsin-Mu Tsai, "Visible light communications for scooter safety," in Proceedings of the 11th Annual International Conference on *Mobile Systems, Applications, and Services*, ACM, pp. 509–510, 2013.

[6] Stefan Schmid, Giorgio Corbellini, Stefan Mangold, and Thomas R. Gross, "LED-to-LED visible light communication networks," in Proceedings of the Fourteenth ACM International Symposium on *Mobile ad hoc Networking and Computing*, ACM, pp. 1–10, 2013.

[7] Nils Ole Tippenhauer, Domenico Giustiniano, and Stefan Mangold, "Toys communicating with LEDs: Enabling toy cars interaction," in *Consumer Communications and Networking Conference (CCNC)*, pp. 48–49, IEEE, 2012.

[8] Stefan Schmid, Giorgio Corbellini, Stefan Mangold, and Thomas R. Gross, "LED-to-LED visible light communication networks," in Proceedings of the Fourteenth ACM International Symposium on *Mobile ad hoc Networking and Computing*, ACM, pp. 1–10, 2013.

[9] Nan Chi, Yuanquan Wang, Yiguang Wang, Xingxing Huang, and Xiaoyuan Lu, "Ultra-high-speed single red-green-blue light-emitting diode-based visible light communication system utilizing advanced modulation formats," *Chinese Optics Letters* **12**, (*1*), 010605, 2014.

[10] Liane Grobe, Anagnostis Paraskevopoulos, Jonas Hilt, *et al.*, "High-speed visible light communication systems," *Communications Magazine*, IEEE **51**, (*12*), 60–66, 2013.

[11] Shlomi Arnon, John Barry, George Karagiannidis, Robert Schober, and Murat Uysal, eds., *Advanced Optical Wireless Communication Systems*, Cambridge University Press, 2012.

[12] Ahmad Helmi Azhar, T. Tran, and Dominic O'Brien, "A gigabit/s indoor wireless transmission using MIMO-OFDM visible-light communications," *Photonics Technology Letters, IEEE* **25**, (*2*), 171–174, 2013.

[13] Zabih Ghassemlooy, Wasiu Popoola, and Sujan Rajbhandari, *Optical Wireless Communications: System and Channel Modelling with Matlab®*, CRC Press, 2012.

[14] Sridhar Rajagopal, Richard D. Roberts, and Sang-Kyu Lim, "IEEE 802.15.7 visible light communication: Modulation schemes and dimming support," *Communications Magazine, IEEE* **50**, (*3*), 72–82, 2012.

[15] Joon-ho Choi, Eun-byeol Cho, Zabih Ghassemlooy, Soeun Kim, and Chung Ghiu Lee, "Visible light communications employing PPM and PWM formats for simultaneous data transmission and dimming," *Optical and Quantum Electronics*, 1–14, 2014.

[16] Shlomi Arnon, "The effect of clock jitter in visible light communication applications," *Journal of Lightwave Technology* **30**, (*21*), 3434–3439, 2012.

[17] IEEE Standard 802.15.7 for local and metropolitan area networks – Part 15.7: Short-range wireless optical communication using visible light.

[18] Chien-Chung Chen and Chester S. Gardner, "Performance of PLL synchronized optical PPM communication systems," *Communications, IEEE Transactions on* **34**, (*10*), 988–994, 1986.

[19] Robert M. Gagliardi, "The effect of timing errors in optical digital systems," *Communications, IEEE Transactions on* **20**, (*2*), 87–93, 1972.

[20] Jin-Wei Shi, Che-Wei Lin, Wei Chen, *et al.*, "Very high-speed GaN-based cyan light emitting diode on patterned sapphire substrate for 1 Gbps plastic optical fiber communication," in *Optical Fiber Communication* Conference, Optical Society of America, 2012, p. JTh2A–18.

[21] Gary Shambat, Bryan Ellis, Arka Majumdar *et al.*, "Ultrafast direct modulation of a single-mode photonic crystal nanocavity light-emitting diode," *Nature Communications* **2**, 539, 2011.

8   DMT modulation for VLC

Klaus-Dieter Langer

8.1   Introduction

Optical free space communication using visible radiation, i.e. light, has been known for a very long time. Some early examples are signaling using fire, the *Heliograph* using sunlight, which is directed to the receiver by means of a mirror, or the *Photophone* invented by Graham Bell (1880). Due to the outstanding success of radio technologies and due to their intrinsic benefits, up to now optical free space communication has remained a niche technology. One of such niche applications took advantage of the immunity from interception, namely the so-called directed transmission for military aims during World War 2 and later on. Different approaches such as optical wireless communication (OWC) using fluorescent tubes are also well recorded in the patent literature but have never achieved a breakthrough.

A revival of this way of wireless communication has come about with the advent of visible light LEDs of increasingly high optical power. While their application initially was limited to signaling (e.g. telltale or warning light), at the turn of the millennium it became apparent that in future lighting would be dominated by LEDs. Henceforth, there has been growing interest in applications using LED-based OWC, or which combine the functions of lighting and optical wireless data transmission. At the same time, the common term *Visible Light Communications* (VLC) was coined for this kind of communication. The major reasons for the steadily rising interest in VLC are the lifetime and improved optical power of white light LEDs particularly, and their progressive adoption, as well as the simplicity of LED modulation via their driving current at a modulation bandwidth in the lower MHz range, see e.g. [1–3]. Moreover, the proliferation of mobile applications using radio frequencies has accentuated concerns about the adequate availability of radio-frequency bands and the limits of transmission capacity in current wireless networks, as well as the related data security issues. In this respect, VLC can offer an additional option for local wireless data links where radio links are not desired or not possible [4–6].

The building blocks of a generic VLC system are shown in Fig. 8.1. Regarding the historical background of VLC and its use for transferring messages via schemes such as the Morse code, it seems to be obvious and straightforward to apply plain on-off keying (OOK). Indeed, simple experimental VLC systems use OOK realized by intensity modulation (IM). At the receiving side, direct detection is applied using a photodetector for optoelectronic signal conversion. While image sensors can be used in

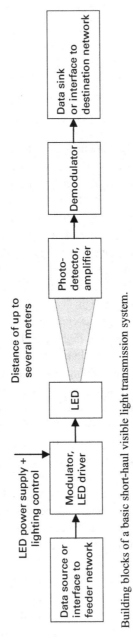

Figure 8.1 Building blocks of a basic short-haul visible light transmission system.

low-speed systems (cf. [3]), high data rates call for a Si-PIN or avalanche photodiode (APD) as the device for optical detection. In such configurations based on OOK and white light LEDs intended for lighting, transmission speeds of e.g. 230 Mbit/s have been achieved [7].

This chapter is focused on indoor applications of VLC, where LEDs are used as an optical source for either (pure) high-speed data transmission or for lighting plus data transmission at bit rates in the upper Mbit/s range (up to Gbit/s speed). Sample applications of high-capacity indoor links are presented in the next section. Given the modulation bandwidth offered by current LEDs, in contrast to the OOK case the targeted data rates require advanced and highly spectrum efficient solutions for modulation, such as discrete multitone (DMT), which is addressed in this chapter. In fact, the highest bit rates demonstrated so far in a single VLC channel use such modulation schemes [8–11].

When considering VLC systems as mentioned before, the reach should be within the range of several meters (say up to ~20 m). Depending on applications such as data broadcast, video streaming or file transfer, either unidirectional or bidirectional links are required in point-to-point (P-t-P) or point-to-multipoint (P-t-MP) configuration. Additional requirements in the dual-use case of LEDs for lighting and data transmission include illumination according to the well-established lighting standards and features without restrictions, no flickering light, dimming, etc., cf. [12]. However, it has to be mentioned that, for instance, the present version of the IEEE standard on VLC only considers OOK, variable pulse position modulation (VPPM), and color shift keying (CSK) as a special scheme with respect to multiple optical sources of different color, while DMT is not yet included [13–15].

The chapter is structured in the following way. We continue with a brief discussion of some typical indoor application scenarios, as they are of interest for the general public and in industrial sectors. The next section is devoted to the relevant characteristics of white light LEDs as the key VLC element. In addition, the optical wireless channel capacity is addressed, as well as LED modulation for exploiting the capacity, and major issues such as the effect of LED non-linearity.

The main part of this chapter presents the DMT modulation scheme and variants thereof, including related signal processing as well as bit and power loading. Substantial effects and consequences such as signal clipping, peak-to-average power ratio (PAPR) and the influence of channel variation are detailed in this section. Subsequently, several recently proposed improvements of the DMT modulation scheme are introduced and various approaches are compared. On that basis, Section 8.6 examines important aspects of system design, implementation issues, and demonstrations in step with actual practice. A summary and an outlook are included at the end of this chapter.

8.2   Indoor application scenarios

VLC can be applied in quite different scenarios. In particular, when high data rates are considered the type of link between transmitter (Tx) and receiver (Rx) is crucial.

According to the mode of propagation of light, there are two generic types of indoor optical wireless links. Firstly the *directed link*, which relies on a non-blocked line-of-sight (LOS) between a highly directed Tx and a narrow field-of-view (FOV) Rx. Secondly, the *diffuse link*, characterized by a wide-beam Tx and a large FOV Rx, where the non-LOS light path relies on numerous signal reflections off the walls and surfaces of objects present in the room [16].

LOS links experience minimal path loss, are rather free from multipath induced signal distortion, and are able to diminish the influence of ambient light. As long as the LOS is not blocked, the link performance only depends on the available power budget. Hence, very high transmission rates are shown to be possible. On the other hand, LOS links require alignment of transceivers and generically provide a very small coverage.

A diffuse link operates entirely without LOS, which results in an increased robustness against shadowing and support of high user mobility within a large coverage area. Thus, a diffuse link scenario enables P-t-MP communication and in general is most desirable from the user's point of view. This is why it evoked much interest from the research community. However, diffuse links suffer from high optical path loss (i.e., they require larger optical powers) and are seriously limited by inter-symbol interference (ISI) because of multipath dispersion, which presents a major degradation factor at higher transmission rates. Beside the power budget, the achievable transmission rate depends also on room characteristics such as size, reflection coefficients of surfaces, etc. It should be mentioned that the effects of multipath fading are negligible in OWC due to square-law detection on a photodiode (PD) of huge size compared to the incoming signal wavelength. This greatly simplifies the link design.

Many VLC applications call for combining the mobility of a diffuse link and the high-speed capability of a LOS link. In order to benefit from the advantages of both connection types, a non-directed LOS (NLOS) link is frequently considered as an alternative. In such a link, LOS and diffuse signal components are simultaneously present at the Rx (assuming a non-blocked LOS path). The equivalent channel response as shown in Fig. 8.2 is characterized by high dynamics in both bandwidth and gain, depending above all, on the LOS prominence. Consequently, the ratio of LOS and diffuse signal components at the Rx, often described by the Rician K-factor, strongly influences achievable data rate,

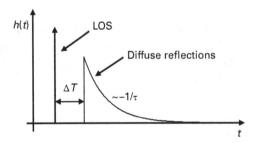

Figure 8.2   Channel impulse response $h(t)$ as a superposition of the LOS and diffuse channel responses, where ΔT indicates the delay between arrival of the LOS signal and the first reflection at the receiver; $1/\tau$ is the decay constant of the diffuse channel.

Table 8.1 Comparison of commonly present link configurations for optical wireless indoor systems.

Link type	Directed LOS	Non-directed LOS	Diffuse
Link rate	highest	high	moderate
Beam pointing	yes	coarse	no
Beam blocking	yes	relaxed	no
User mobility	low	medium	high
Dispersion (multipath)	none	medium	high
Path loss	low	extended	high
Impact of ambient noise	low	medium	high

impact of ISI, and ambient light, while the Tx-Rx alignment is relaxed and P-t-MP communication is possible.

Table 8.1 summarizes important features of the basic Tx-Rx configurations in OWC. Because in addition to the link type the lighting scenario may also vary while used for VLC, a dynamic data rate adaptation appears necessary, as already proposed in [17] and [18]. Such a feature would enable efficient use of the channel and its instantaneous character, and thus would substantially contribute to making VLC links robust and ready for various use cases. There are several approaches and options based on the fundamental principles of high-speed VLC transmission, which are addressed in this chapter. Thus, we will not dive deeper into the matter of dynamic data rate adaptation. In the following, some illustrative VLC scenarios, mainly using high-speed transmission, are discussed.

Pure wireless broadcast links could provide passenger information, etc., via the general lighting, e.g. on underground stations (Fig. 8.3) or inside the metro cars, where lights are always on anyway. Such indoor systems require unidirectional data transfer (streaming), a few meters transmission reach and a wide FOV, which is inherently given by lighting. The basic functionality of such an application is quite similar to the traditional and most commonly used optical wireless application (at very low speed, but using infrared light), namely the remote control. However, today's technology can provide very high data rates up to the Gbit/s range.

Another example of VLC combined with general lighting, as shown in Fig. 8.4, is also known as optical WiFi or Li-Fi (light fidelity). In wireless local area network (WLAN) scenarios like this, the downlink is provided in the same way as above, while the uplink from the laptop to an access point at the ceiling can be established, e.g. using a LOS infrared (IR) link (see e.g. [19]). Such a system can offer bidirectional communication with wide FOV P-t-MP visible light downlinks at speeds in the range of several Mbit/s or much more depending on the link conditions (LOS or diffuse), and IR directed LOS P-t-P uplinks of a few Mbit/s, assuming both fair Tx-Rx alignment and optical power [20].

Machine-to-machine communication is expected to be a further broad field of VLC application, e.g. with respect to wireless exchange of high data volumes within small and dense communication cells. It is also assumed that in industrial environments there may be harsh electromagnetic conditions going along with the highest demands on security

Figure 8.3   Broadcast of multimedia passenger information by visible light on an underground station.

Figure 8.4   Optical wireless LAN scenario where the downlink is provided by VLC while the uplink is realized, for example, by infrared light.

Figure 8.5   Bidirectional VLC applied in a production line, e.g. in order to trigger the device under assembly for performance tests and to receive the results during shift on the conveyor.

and reliability making it difficult or even impossible to use radio-frequency systems. VLC directed LOS links on the other hand could provide suitable high-speed bidirectional links. As an example, Fig. 8.5 illustrates a scenario where performance tests of products under assembly happen while the units are moving on a conveyor, and data exchange with an assessing central server is required at the same time.

8.3   Aspects of high-speed VLC transmission

In particular, when combined with lighting, key enablers for VLC are that the LEDs do not flicker, that only a minimum of extra power is needed, that VLC works at the commonly used brightness levels, and that well-established functions of lighting such as dimming are possible without reservation [13, 14, 21]. These aspects are addressed in Section 8.6.1. From the transmission technology point of view, the LED modulation bandwidth is just as crucial as the wireless channel capacity and how to exploit it efficiently. These subjects will be discussed in the subsequent sections.

8.3.1   LED modulation bandwidth

White light LEDs, as key VLC elements, are manufactured by appropriately adding the light of three or four colored emitters, i.e. red, green and blue (RGB, trichromatic) or red, yellow, green and blue (RYGB, tetrachromatic) LED devices, respectively. As an alternative, a single blue emitter chip coated with a yellowish phosphor layer (YB, dichromatic) is used. The RGB and RYGB white light sources provide the desired spectral output, but are hardware-intensive as multiple LEDs are required. In addition, they tend to render pastel colors unnaturally, a fact which is largely responsible for the poor color-rendering index of RGB white light. As a result, the YB-LEDs are currently the device of choice for illumination, and thus also for low-cost VLC.

While typical RGB and RYGB LEDs created for lighting purposes offer a 3 dB modulation bandwidth of some ten MHz, the bandwidth of YB-LEDs is much lower, i.e. in the order of a few MHz, see e.g. [22, 23]. This is because of the slow temporal response of the yellow phosphor layer. On the other hand, as explained in [24], a bandwidth of several tens of MHz could be expected in the absence of the phosphor layer. Accordingly, it is good practice to place a blue optical filter in front of the PD at the Rx side in order to receive only the blue component of the light and to take advantage of its larger modulation bandwidth [25]. This, however, comes along with a lower power budget and signal-to-noise ratio (SNR), as the major portion of the received light spectrum is filtered out [26]. Alternatively, equalization techniques can be used to combat the influence of the phosphor layer [27, 28]. To cite an example, an array of 16 YB-LEDs was modified in that way to have a bandwidth of 25 MHz without blue filtering [29].

Regarding the LED modulation bandwidth, in a more general sense it is worth mentioning that according to experimental studies the modulation bandwidth of white light as well as of colored LEDs can be exploited far beyond the 3 dB drop. For instance, in [30] a bandwidth of 100 MHz has been used for modulation, while the LED 3 dB bandwidth was about 35 MHz.

Because, in addition to the commonly used (inorganic) LEDs in lighting systems, organic LEDs (OLEDs) also become attractive for replacing large area luminous sources, it should be noted that OLEDs are considered in VLC too. However, they offer an inherently low modulation bandwidth in the range of about 100 kHz [31, 32], and thus they are not in the scope of this chapter.

8.3.2   Channel capacity

As briefly discussed in the context of VLC applications above and summarized qualitatively in Table 8.1, the capacity of the optical wireless channel in a given indoor scenario strongly depends on the presence of both LOS and diffuse signal components. A useful means of describing the channel state by a single parameter is the Rician K-factor, which is defined as the ratio of the (electrical) LOS and diffuse channel gain (loss), i.e. K [dB] $= 20 \log(\eta_{LOS} / \eta_{DIFF})$. Following the analysis in [33, 34], where an empty model room and realistic parameters are assumed as an example, the frequency response is calculated and illustrated in Fig. 8.6 (a). The diagram shows the composite channel frequency response magnitude obtained from the analytical model for several illustrative K-factors. Clearly, the channel response highly depends on the LOS prominence (described by the K-factor). Where the LOS is weak or blocked, the response is approximately low-pass and the bandwidth is quite poor. As the LOS gets more pronounced, the channel response varies until it becomes almost flat for sufficiently large K-factors, rendering bandwidths up to an order of magnitude greater than in the diffuse case. The notches in the channel characteristic are due to destructive interference of the two frequency components. Assuming, e.g. a realistic working area hot-spot scenario, the K-factor span of interest is about −20 to +25 dB.

The simplest way to obtain reliable connectivity under all channel conditions within the coverage area of a given scenario is to realize a statically designed system, with transmission parameters fixed according to the worst case. However, such a system would not be able to benefit from the channel properties. In order to illustrate the potential of the indoor optical wireless channel, the upper capacity bound in terms of transmission rate for different channel states and two optical power limits ($P_O = 0.1$ and 0.4 W) is presented in Fig. 8.6 (b). For comparison, the curves of a statically designed system, aimed to guarantee a constant transmission performance in the whole area of interest are also shown. From Fig. 8.6 (b) it becomes clear that it is extremely attractive to exploit the channel capacity, in particular at growing K-factors, where the link becomes more and more transparent. The gain (with respect to a worst-case design) grows with the K-factor, and also by increasing the optical transmit power, where it becomes significant even at decreasing values of K. The information rate also depends on the electrical bandwidth, which may be limited by the LED as outlined above in Section 8.3.1.

The calculations were made when the transmit power was allocated as best possible to the subcarriers of a multiple subcarrier modulation (MSM) scheme. In doing so it is also shown in [35] that optimum power allocation gives negligible advance when the electrical bandwidth is limited to about 20 MHz, but offers more improvement at higher bandwidths, particularly for low Tx powers under MSM methods, which are the focus of this chapter. The topic of Tx power dynamic range and its influence on the information rate is analyzed in [36]. It is shown there that a Tx with a wide linear dynamic range of 20 dB or more provides sufficient electrical power for OWC with an optical transmit power close to the boundaries of the dynamic range, where the LED appears to be off or powered close to its maximum. It is also shown there that an average optical power sweep over 50% of the dynamic range can be accommodated using an appropriate modulation scheme, if the information rate is reduced by only ~10%.

In addition to the theoretical work discussed above, results of an experimental channel characterization can be found for example in [37], where a VLC channel bandwidth of 63 MHz is reported for a certain room geometry.

The cited work and further results indicate that although the well-known Shannon channel capacity formula for a band-limited, average power-limited additive white Gaussian noise (AWGN) channel cannot be applied directly to the VLC channel, it can

(a)

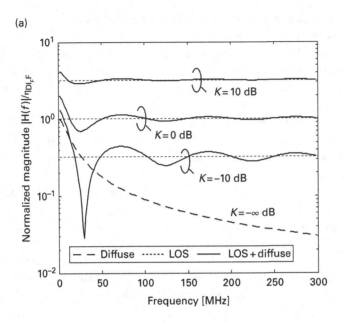

(b)

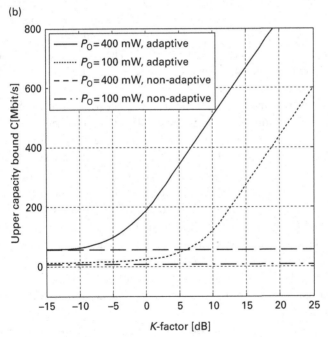

Figure 8.6   (a) Frequency response of the optical wireless channel in a model room for different K-factor values. The derivation is outlined in [33], where $\Delta t = 10$ ns delay between LOS and diffuse signal components is assumed. (b) Upper capacity bound C of an adaptive and a non-adaptive OWC system (bandwidth 100 MHz) as a function of the channel state and the mean optical power P_O as parameter.

be adapted in many cases to find the limit of reliable OWC. As a conclusion, it is noted that in non-diffuse situations the channel presents much more capacity than offered by the LED, depending on the Rician K-factor. This calls for advanced spectrum efficient LED modulation if high-speed communication is desired. In order to maximize the system throughput, while reliable operation and full coverage are maintained, an OWC system could be designed that is bandwidth- and rate-adaptive. Such an adaptive system reduces the data rate under adverse channel conditions until a desired bit error performance is attained. However, such a feature requires a reliable low-speed feedback link for transferring the necessary channel state information from the Rx to the Tx.

8.3.3 Considerations on high-speed LED modulation

With the aim of developing simple and low-cost VLC systems, it is an obvious step to choose simple and well-known modulation formats such as OOK, pulse-width modulation (PWM), or M-ary pulse-amplitude modulation (M-PAM), which can be applied in a straightforward way using intensity modulation and direct detection. The literature provides a lot of proposals and demonstrations on a low-speed level, but also some examples where several 100 Mbit/s have been achieved using OOK, cf. for instance [7, 38]. However, if higher transmission speeds are needed, the above mentioned modulation schemes begin to suffer from the undesired effects of ISI due to the non-flat frequency response of the optical wireless channel [15]. Hence, a more resilient and, in view of the limited LED bandwidth, a more spectrum efficient technique such as MSM is required.

In the VLC transmission chain, the LED device is a major source of non-linear distortion. This is due to a non-linear transfer function within the LED's operating range. As the significance of such behavior strongly depends on the modulation scheme used, non-linearity aspects have been extensively studied in numerous papers, such as [39–43]. The LED non-linearity is of particular importance if spectrally efficient MSM modulation is applied in order to exploit the LED's bandwidth. Using such modulation schemes, which this chapter is focused on, the dynamic range and linearity of the emitter device may severely limit the achievable performance [4]. In order to mitigate such restrictions the selection of a proper modulation format is crucial [44], [45], while the LED operating point has to be defined carefully [46], and signal clipping prior to modulation has to be considered [47].

Beyond the non-linearity issue, further effects on the LED performance, such as LED degradation due to ageing or junction temperature variation have to be considered. First studies on this subject have been published, e.g. in [48], however, further investigations on the LED behavior under VLC operation conditions are necessary.

8.4 DMT modulation and variants

MSM techniques are modulation schemes where information is modulated onto orthogonal subcarriers located in the frequency band considered. The sum of the modulated

Table 8.2 Contrast between typical OFDM systems and optical multimode transmission based on intensity modulation and direct detection.

Area of application	Typical example	Carrier of information	Type of detection	Special receiver requirement
Electrical OFDM transmission	Radio transmission, e.g. WiFi, DVB-T	Electrical field	Coherent	Local oscillator
Optical OFDM transmission	Long-haul high-speed single mode fiber (SMF) links	Optical field	Coherent (only one optical mode)	Local oscillator
Optical IM/DD transmission	Multi-mode fiber (MMF, POF) links, OWC	Optical intensity	Direct (multiple optical modes)	None

subcarriers is then modulated onto the instantaneous power of the transmitter. Usually MSM is implemented by *orthogonal frequency division multiplexing* (OFDM), which has been widely employed in both wired and wireless digital communications. Applications of the basic OFDM principle include WLAN and terrestrial digital video broadcasting (DVB-T), as well as digital subscriber line systems (xDSL) and power line communication (PLC), where a baseband version of OFDM, better known as *discrete multitone* (DMT) modulation is used. Due to the capability to mitigate ISI, and other advantages including the ability to adapt easily to different channels, MSM was also considered for OWC [49].

Conventional systems such as WLAN or DVB-T use coherent transmission, where the signals are in general complex and bipolar. The same applies to laser-based long-haul optical transmission systems using single-mode fibers. Concerning *optical multimode transmission* as in VLC, it is, however, extremely difficult to collect substantial signal power at the Rx in a single electromagnetic mode. In the final analysis, this means that *intensity modulation with direct detection* (IM/DD), as also used for instance in OOK and PWM-based systems, has to be considered as the only practicable transmission method. Thus, only the light intensity (and not the phase) represents the information to be transmitted, i.e., the transmitted signal is of real and non-negative (unipolar) value. At the receiving end, a photodetector produces an electrical current proportional to the received power, and accordingly proportional to the square of the received electrical field [16]. The differences between traditional OFDM transmission using coherent detection and IM/DD systems are briefly summarized in Table 8.2.

In consequence of IM/DD transmission, the conventional OFDM modulation scheme cannot be directly applied in optical multimode systems. Therefore, researchers have devoted significant efforts to designing OFDM-based modulation schemes, which are purely unipolar. Using OFDM in VLC was first proposed by Tanaka *et al.*, and their basic studies can be found in [50].

The task to create a unipolar signal for transmission is commonly solved by adding a DC bias to the bipolar OFDM signal [18, 25]. We call this scheme *DC-biased DMT*, however, it is also known as DC-biased OFDM and DC-clipped OFDM. Another method clips the entire negative excursion of the OFDM waveform. Impairments from clipping noise are avoided by appropriately choosing the subcarrier frequencies for modulation. This technique is called *asymmetrically clipped optical OFDM* (ACO-OFDM) [51]. A third method also clips the entire negative signal excursion, but modulates only the imaginary parts of the subcarriers such that the clipping noise becomes orthogonal to the desired signal. This technique is well-known as *pulse-amplitude-modulated discrete multitone modulation* (PAM-DMT) [52]. These modulation schemes and their properties when applied to VLC are discussed in the subsequent sections.

8.4.1   DC-biased DMT

DMT as a baseband version of OFDM modulation is a key technique used on slowly time-varying two-way channels, e.g. in copper-based xDSL and PLC systems. It is important to know that DMT is also of relevance for low-cost short-range optical transmission using multimode silica fibers and plastic optical fibers [52, 53]. Usually, modulation onto multiple subcarriers of different frequencies for simultaneous transmission, as well as demodulation, are based on discrete Fourier transformation (DFT) [54, 55]. Thus, inverse fast Fourier transform (IFFT) and fast Fourier transform (FFT) are the main building blocks of the Tx and Rx, respectively.

On the Tx side, first of all the serial data stream is partitioned into multiple parallel streams of lower data rates, typically followed by mapping to a quadrature amplitude modulation (QAM) constellation, as illustrated in Fig. 8.7. A straightforward way to obtain real-valued time domain signals, as required in LED-based VLC, is to use the well-known property that the N-DFT of a real-valued sequence has conjugated symmetric coefficients around the point $N/2$ [56]. This means that by enforcing conjugate symmetry (often referred to as *Hermitian symmetry*) on the IFFT input vector [**X**] in the frequency domain, a real-valued time domain signal can be directly obtained at the output of the IFFT block.

Following this approach, with $N/2$ (actually $(N/2) - 1)$) independent subcarriers envisaged in the system to carry information, an N-IFFT block is required to generate a real-valued OFDM/DMT symbol. According to the Cooley–Tukey algorithm [56], the complexity of the FFT operation scales with $N\log_2 N$. Hence, as a moderate number of subcarriers is considered in optical wireless systems, the size of the DFT blocks is not an implementation issue. On the other hand, DFT enables digital implementation of DMT modulation without prohibitive analog filter banks.

The IFFT input vector $\mathbf{X} = [X_0\ X_1\ \cdots\ X_{N-1}]^T$ consists of the data to be transmitted (elements $X_1, X_2, \cdots, X_{(N/2)-1}$) and the further elements defined according to the Hermitian symmetry constraint as

$$X_n = X_{N-n}^*, \quad \text{for } 0 < n < N/2, \quad X_0 \in \mathbb{R}, \quad X_{N/2} = 0. \tag{8.1}$$

The first input X_0, corresponding to zero frequency, must be real-valued and is generally left unmodulated. It can be set to zero or alternatively to the DC level of the output signal (see below).

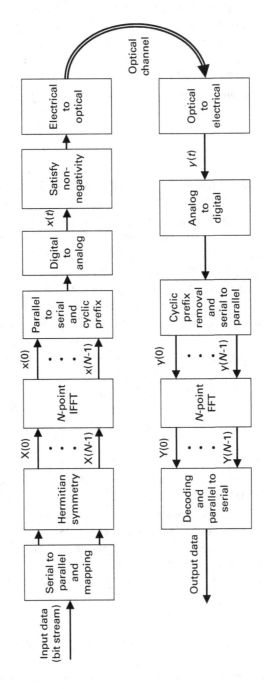

Figure 8.7 Building blocks of DMT-based transmission over an optical IM/DD channel. Note that the tasks of satisfying non-negativity and digital-to-analog conversion may be swapped depending on the hardware implementation of the DC biasing.

After creation of the input vector, [**X**] is fed into the N-point IFFT block, as shown in Fig. 8.7. Assuming data symbols of complex-valued modulation formats (usually M-QAM), where the input vector elements appear in the form $X_n = a_n + jb_n$, the time samples at the N-IFFT output are

$$
\begin{aligned}
x(k) &= \frac{1}{N} \sum_{n=0}^{N-1} X_n \, \mathrm{e}^{\mathrm{j}2\pi nk/N} = \frac{X_0}{N} + \frac{1}{N} \sum_{n=1}^{(N/2)-1} X_n \, \mathrm{e}^{\mathrm{j}2\pi nk/N} + \frac{1}{N} \sum_{n=1}^{(N/2)-1} X_n^* \, \mathrm{e}^{-\mathrm{j}2\pi nk/N} \\
&= \frac{X_0}{N} + \frac{1}{N/2} \sum_{n=1}^{(N/2)-1} \mathfrak{Re}\left\{ X_n \, \mathrm{e}^{\mathrm{j}2\pi nk/N} \right\} \\
&= \frac{X_0}{N} + \frac{1}{N/2} \sum_{n=1}^{(N/2)-1} a_n \cos\left(2\pi nk/N\right) - b_n \sin\left(2\pi nk/N\right) \\
&= \frac{X_0}{N} + \frac{1}{N/2} \sum_{n=1}^{(N/2)-1} \sqrt{a_n^2 + b_n^2} \, \cos\left(2\pi nk/N + \arctan(b_n/a_n)\right),
\end{aligned}
\tag{8.2}
$$

where $k = 0, 1, \ldots, N-1$ denotes the index of the time domain sample. Apart from the conjugate-symmetry property of the input vector given by Eq. (8.1), the symmetry property of the DFT twiddle factors $\mathrm{e}^{\mathrm{j}2\pi(N-k)k/N} = \mathrm{e}^{-\mathrm{j}2\pi Nk/N}$ has also been exploited in Eq. (8.2). Obviously, a sum of $(N/2)-1$ sampled real-valued cosinusoids is obtained at the IFFT output.

In order to mitigate effects of the multipath channel and to avoid ISI, OFDM and DMT use a guard interval, which is placed between the transmitted symbols (blocks) in the time domain. This guard band is formed simply by taking a number of L samples from the end of each symbol, and copying them as its prefix, as shown in Fig. 8.8. It is hence referred to as *cyclic prefix* (CP). This $(M + L)$-point sequence corresponds to the samples of the multicarrier DMT time-discrete sequence to be transmitted, which is referred to as a DMT symbol.

In order to receive and properly demodulate the DMT symbols, two conditions have to be satisfied. Firstly, the length of the DMT symbol without CP should be longer than or equal to the duration of channel impulse response $h(t)$ in order to avoid ISI. Additionally, the CP length should be chosen so that its duration is longer than or equal to the delay

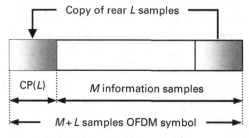

Figure 8.8   Generation of the cyclic prefix guard interval and structure of a DMT symbol. CP: cyclic prefix.

spread of $h(t)$. Although the CP introduces some redundancy, and thus reduces the overall data rate, it eliminates both ISI and inter-carrier interference from the received signal and is the key to simple equalization in OFDM [55].

When assuming a limited signal bandwidth B in the baseband, which is divided among $N/2$ independent subcarriers, the subcarrier spacing is $\Delta f = 2B/N$, while the frequencies in the N-IFFT block cover the bandwidth of $2B$ because of the Hermitian symmetry constraint. The period of a DMT symbol, containing N time samples, is $T_{FFT} = 1/\Delta f$, which leads to the sample interval $T_{sam} = 1/2B$ in the time domain. According to the sampling theorem, this is sufficient to completely determine the continuous signal at the output of the digital-to-analog (D/A) converter. As the CP of length $T_{CP} = LT_{sam}$ has to be added after the IFFT, the actual duration of the real-valued DMT symbol is $T_{DMT} = T_{CP} + T_{FFT}$. Considering in addition Eq. (8.2) and the identity

$$2\pi \frac{nk}{N} = 2\pi n \frac{2B}{N} \frac{k}{2B} = 2\pi n \Delta f k T_{sam} = 2\pi f_n t_k, \qquad (8.3)$$

where $f_n = n\Delta f$, $n = 1, 2, \ldots, N-1$ and $t_k = kT_{sam}$ for $k = 0, 1, \ldots, N-1$, the continuous-time signal after D/A conversion can be expressed as

$$x(t) = \frac{X_0}{N} + \frac{1}{N/2} \sum_{n=1}^{(N/2)-1} A_n \cos(2\pi f_n t + \varphi_n), \quad -T_{CP} \le t < T_{FFT}. \qquad (8.4)$$

In this expression, $A_n = \sqrt{a_n^2 + b_n^2}$ and $\varphi_n = \arctan(b_n/a_n)$ are the amplitude and the initial phase of each cosinusoid on the frequency occupied by the nth subcarrier, determined by the amplitude and phase of the complex-valued symbol X_n at the corresponding IFFT input. Note that the signal $x(t)$ could also be obtained with a bank of $(N/2)-1$ analog filters. After D/A conversion, an analog low-pass filter is implemented to suppress the aliasing spectra. At the same time, this filter performs interpolation of the discrete signal waveform.

Figures 8.9 and 8.10 illustrate an example of the signals at the input and output of the DMT modulator. In this example, three out of $N = 16$ orthogonal carriers in a bandwidth of $B = 20$ MHz are modulated by 16-QAM to generate an input vector $[\mathbf{X}] = [0, (1+j), (3-j), 0, (-3+3j), 0, 0, 0]^T$ (before Hermitian symmetry enforcement). The input of the 16-IFFT block is shown in Fig. 8.9, while the contributions of individual subcarriers at the output as well as the resulting output signal are illustrated in Fig. 8.10 (a)–(c) and (d) respectively.

If the bias was not introduced directly at the system input by setting X_0 appropriately, i.e., if $X_0 = 0$ was chosen, non-negativity of the transmit signal $x(t)$ has to be achieved after D/A conversion. A very common and simple method is to add a fixed DC bias to the bipolar DMT signal, as illustrated in Fig. 8.10 (e), cf. for instance [25, 49, 52, 57]. The required DC bias is equal to the maximum negative amplitude of the DMT signal. Owing to the high PAPR of DMT signals, a bias of at least twice the standard deviation of the bipolar DMT signal distribution is required to minimize clipping [58]. Any remaining negative values would be clipped, which also applies to positive peaks upon exceeding the limiting amplitude. Such symmetrical or asymmetrical clipping

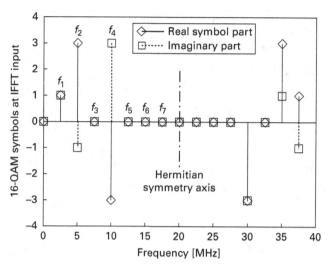

Figure 8.9   Example of 16-QAM symbols on individual subcarriers in f-domain at the IFFT input according to the input vector $[\mathbf{X}] = [0, (1 + j), (3 - j), 0, (-3 + 3j), 0, 0, 0]^T$ before Hermitian symmetry enforcement. Further parameters are $N = 16$, $B = 20$ MHz, and $L = 2$.

introduces clipping noise and thus affects transmission. In practice, controlled clipping of high negative peaks in front of the optical source is often permitted, given an unlikely occurrence in a DMT signal. Then, as long as the effect of clipping on the link performance is tolerable, the required DC component can be reduced to a certain extent [33], cf. also Section 8.6.1. An optimal DC bias of a symmetrically clipped signal can be inferred, e.g. from [59, 60]. It is shown there that iterative decoding with clipping noise estimation and subtraction can reduce the bit error ratio (BER) at the expense of an increased computational complexity.

Returning to Fig. 8.7, the signal reaches the Rx after undergoing the influences of the time-dispersive channel and ambient light as the dominant source of noise in the channel. A simple (low-cost) photodetector is used to convert the IM signal to an electrical signal, while the bias component is discarded by AC-coupling. After analog-to-digital (A/D) conversion, the CP is removed and N-point FFT processing is performed. Thanks to the preserved orthogonality, the subcarriers can be processed separately. In principle, only the outputs $n = 1, 2, \ldots, (N/2) - 1$ need to be further regarded, since they are the ones carrying information. Because the frequency response over the subcarrier bandwidth can be considered as flat fading, the signal is usually equalized simply by means of a single-tap linear feed-forward equalizer with zero forcing or minimum mean squared error (MMSE) criteria before decoding.

Adding the DC bias at the Tx results in extra power consumption to an equivalent extent. If the light source is used at the same time for illumination, this amount of power fulfills its purpose with respect to lighting and is thus not wasted. Only if illumination is not required, such as in the uplink of a Li-Fi system, the DC bias can significantly jeopardize energy efficiency [15].

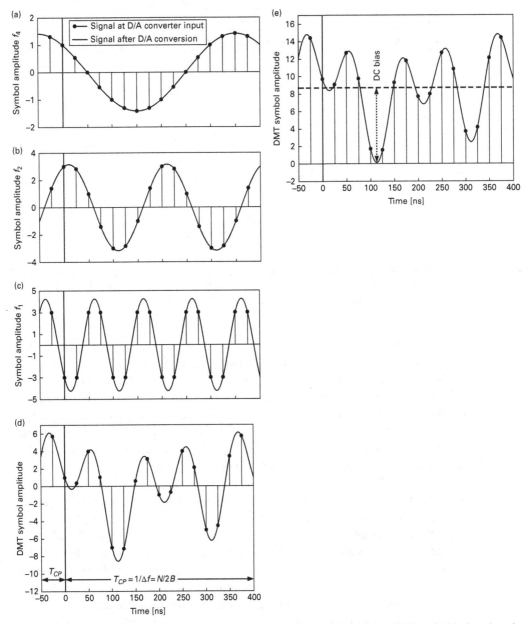

Figure 8.10   (a)–(c) The individual contributions of modulated subcarriers; (d) the DMT symbol in the t-domain. Both sampled and continuous signals before and after D/A conversion are shown. Note that the amplitudes are scaled by N to achieve the same level as in the f-domain. In this example, the real-valued bipolar DMT signal with $N = 16$ subcarriers used in the example (d) is made unipolar by adding an appropriate DC bias (e).

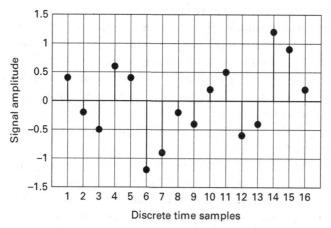

Figure 8.11   Example of an ACO-OFDM time domain signal with $N = 16$ subcarriers (the CP is not considered here). The positive and negative parts of the waveform are anti-symmetrical along the zero-axis. Thus, hard-clipping the negative values at the zero level only will discard redundant information.

8.4.2   Asymmetrically clipped optical OFDM (ACO-OFDM)

The main objective of improving the preparation of OFDM signals for IM/DD transmission is to reduce the DC bias, which is indispensable in the DC-biased DMT scheme, to a minimum value. Note that the LED turn-on voltage in fact defines the lower bound of the bias. For the sake of simplicity this parameter is neglected in the following. Approaches for improving the power efficiency are driven by the basic aim to create a real-valued bipolar time domain signal, where the entire information is present at least in the positive parts. This would enable simply clipping the negative parts of the signal when modulating the optical source. The main embodiment of such asymmetric clipping is the method of ACO-OFDM introduced by Armstrong *et al.* [51, 58].

While the basic function blocks are the same as in DC-biased DMT, the ACO-OFDM scheme exploits the properties of Fourier transform by introducing a constraint on the subcarriers used for modulation, i.e., only odd-numbered subcarriers are used whereas the even subcarriers are set to zero. Considering again Hermitian symmetry, up to $N/4$ subcarriers (CP included) are used for modulation [51, 61]. As illustrated in Fig. 8.11, the IFFT output presents time samples along the zero-axis, where the negative part is an anti-symmetrical copy of the positive part. Finally, this means that the information of both parts is the same. For that reason, clipping of the negative part will provide a unipolar signal without any loss of information. Related proofs can be found in [43, 51].

Insertion of a CP and D/A conversion is done in the same way as in the case of DC-biased DMT. Subsequently, the unipolar signal is used for modulating the LED.

The idea of using only the odd-numbered subcarriers for data transmission includes the fact that all of the intermodulation products resulting from the asymmetric clipping are orthogonal to the data, i.e., clipping distortion falls on the unused even-numbered subcarriers and does not affect the odd-numbered ones. Even though the effect of clipping does not result in inter-carrier interference on the data-carrying odd subcarriers, it reduces all their amplitudes by half [51, 58, 62]. Moreover, the restriction to use only odd-numbered subcarriers can degrade performance in a frequency-selective channel. Particularly when bit and power loading is employed, where the allocation of bits and power per subcarrier is adapted to the SNR of the frequency-selective channel [54, 63], the subcarrier constraint will result in a non-optimal adaptation to the channel response and thus in an unfavorable system performance. This calls for carefully controlled and effectively compensated solutions [64], in particular if the number of subcarriers is small.

At the Rx, the PD detects the transmitted intensity and the analog signal is converted into a digital one. It is well-known that OFDM systems are highly sensitive to synchronization errors. Due to the fundamentally different waveforms in ACO-OFDM and traditional OFDM systems, conventional synchronization could fail if applied directly. A technique tailored to the ACO-OFDM scheme using an appropriate training sequence is presented in [65]. According to [66], the loss of bandwidth efficiency associated with a training sequence can be avoided in systems employing ACO-OFDM if the CP-based method is replaced by the presented technique of low-complexity blind timing synchronization.

In conclusion, ACO-OFDM has a significantly lower detrimental DC component compared to DC-biased DMT. Thus, the modulation scheme is highly power efficient, but at the expense of exploiting only half of the subcarriers for data transmission.

8.4.3   Pulse-amplitude-modulated discrete multitone (PAM-DMT)

The concept of pulse-amplitude-modulated DMT (PAM-DMT) introduced in [52] also aims at asymmetric clipping at zero value and transmission of only the positive parts of the DMT signal as in ACO-OFDM. This means that again anti-symmetry of the time domain signal is needed after IFFT.

Firstly, as in DC-biased DMT and ACO-OFDM, Hermitian symmetry is induced in order to achieve real-valued signals in the time domain. However, only the imaginary components of the PAM input signal are used further, while the real components are set to zero. The IFFT then presents real-valued time domain samples that exhibit anti-symmetry, as desired. Similarly to DC-biased DMT, up to $N / 2$ subcarriers (including CP) are used, i.e. the elements $(X_1, X_2, \cdots, X_{(N / 2) - 1})$ of $[\mathbf{X}]$ carry the imaginary PAM signal components, while $X_0 = 0$. After CP insertion and along with D/A conversion, the entire negative excursion of the electrical waveform is clipped without any loss of information, and is then used for driving the LED.

In [52] it is shown that the distortion resulting from asymmetric clipping falls orthogonally onto the real-valued parts of the PAM signal, without influencing the imaginary parts modulated with information, and thus without affecting the system performance. Proofs of anti-symmetry and the property of the clipping distortion terms can be found in [43]. At the Rx, demodulation and decoding the data are performed straightforwardly and similarly to the schemes discussed before.

In conclusion, PAM-DMT uses all subcarriers as in the DC-biased DMT scheme, but the modulation is restricted to just one dimension. Hence, PAM-DMT has the same spectral efficiency as ACO-OFDM. Regarding the power efficiency, the PAM-DMT characteristics are similar to those of the ACO-OFDM scheme.

8.4.4   DMT/OFDM performance and mitigation of disruptive effects

Numerous analytical studies on characteristics of the modulation schemes described above and on differences in their performance in VLC have been published in recent years. The most important items are discussed in this section.

If reduction of the high PAPR is required, simple non-linear techniques can be applied, possibly in combination with some form of predistortion [55, 67]. Various precoding techniques for PAPR reduction in asymmetrical clipped systems such as ACO-OFDM and PAM-DMT are analyzed in [68]. It was shown there that precoding with small effort could gain PAPR reduction of about 3 dB, making this technique attractive for PAPR reduction in systems with asymmetrical clipping.

Several comparative analyses have been performed including mathematical description of effects and simulations under various conditions, cf. in particular [69–73]. To shed light on one of the results, it was found that ACO-OFDM and PAM-DMT have virtually identical performance at different bit rates and spectral efficiencies, as demonstrated in the study [70], and is also indicated in several other publications. This is because in ACO-OFDM, half of the subcarriers are filled, but in PAM-DMT half of the quadrature is filled. Therefore, the same performance is obtained when a flat frequency response is considered.

In practice, clipping the LED modulation signal clearly affects the system performance. The modulation order and other parameters of relevance such as the bias level (including the LED turn-on threshold) are therefore of high significance. While AWGN is the main type of noise in the case of low SNR, clipping distortion dominates at large SNR values. Thus, LED clipping effects have to be considered and the modulation order as well as the power of the transmit signal should be optimized. Methods for optimization are given e.g. in [45]. Another analytical study on the trade-off between biasing and clipping suggests that rather than eliminating all clippings, the SNR is in fact optimal with some deliberately introduced non-linear clipping distortion because of higher power efficiency at a lower bias level. A corresponding biasing strategy is proposed in [74]. This method does not require any on-line calculation and can be used in different optical channels with different Rx figures and modulation schemes; however, an uplink between Rx and Tx is necessary.

Besides asymmetric clipping to achieve non-negativity, the signal for LED modulation often has to be double-sided clipped in order to fit into the linear range of the LED [45, 73]. This introduces non-linear distortions. In a more advanced approach scaling of the transmit signal is proposed in order to accommodate the large PAPR in the LED's linear range and to adapt the DC bias accordingly. More precisely, such signal shaping as presented in [71] conditions the transmit signal to meet the optical power constraints of the Tx by means of scaling (realized by digital signal processing, DSP) and by DC biasing (realized in the analog Tx circuitry). An optimal signal shaping enables the Gaussian transmission signals to minimize the electrical SNR requirement. As in the case

Table 8.3 Computational complexity needed for the three modulation schemes at low (upper three rows) and high (lower three rows) electrical bandwidth, expressed in terms of real operations per bit. For parameters in detail, cf. [70].

Bit rate (Mbit/s)	Electrical bandwidth (MHz)	DC-biased DMT		ACO-OFDM		PAM-DMT	
		Tx	Rx	Tx	Rx	Tx	Rx
50	25	43.0	45.9	48.5	50.0	48.5	51.5
100	50	43.0	45.9	48.5	50.0	48.5	51.5
300	150	42.4	45.3	48.1	49.6	48.1	51.1
50	50	86.1	91.9	97.0	100.0	97.0	102.9
100	100	85.4	91.2	96.7	99.6	96.7	102.6
300	300	95.3	101.1	106.5	109.4	106.5	112.4

of double-sided clipping, shaping of the Gaussian time domain signals in ACO-OFDM and DC-biased DMT results in a non-linear signal distortion, which is precisely analyzed in [75]. This analysis can be used to translate the signal scaling and DC biasing for a given dynamic range of the LED into electrical SNR, and therefore to BER performance.

As a summary of the numerous analyses and comparisons, both modulation formats using asymmetrical clipping, i.e., ACO-OFDM and PAM-DMT, are best at low spectral and high power efficiency [69, 70, 72]. It is also important to note that multipath dispersion could break the symmetry at the zero line for both of these schemes [52]. Since the symbol period usually by far exceeds the value of dispersion, this effect should not represent a serious issue; however, it has to be confirmed by practical experience. If modulation at higher spectral efficiency is needed, DC-biased DMT performs closer to ACO-OFDM. This is because the clipping noise penalty for DC-biased DMT becomes less significant than the penalty for the larger constellations required if otherwise the ACO-OFDM or the PAM-DMT scheme is used. Thus, DC-biased DMT is expected to deliver the highest throughput in applications where the additional DC bias power required to create a non-negative signal does not matter or can serve a complementary functionality, such as illumination. It should be mentioned that the underlying comparisons typically assume a perfectly compensated LED non-linearity, e.g. using a suitable predistortion. Hence, practical verification is also necessary in that respect.

Beyond performance aspects, the computational complexity of the modulation schemes discussed is of importance. A comparative study on the computational complexity can be found in [70]. The same number of actually used subcarriers in DC-biased DMT and ACO-OFDM was chosen there in order to make a fair comparison. For PAM-DMT, since only one dimension is used to transmit data, the number of subcarriers used was twice that for the other schemes. In other words, from the DFT point of view, the DFT size of the DC-biased DMT scheme is half that of the further schemes.

Table 8.3 summarizes the computational effort needed in the three cases at different bit rates and electrical modulation bandwidths. As can be seen, the complexity of ACO-OFDM and PAM-DMT is nearly the same, while DC-biased DMT has the lowest computational complexity among the three formats, due to the smaller DFT size.

Theoretical work such as mathematical analyses and simulations including ray tracing has to be verified and complemented by experiments under conditions close to reality. At this stage of development in the case of modulation schemes such as DMT, there is the difficulty of time-consuming hardware implementation including DSP algorithms. As an alternative, the widely recognized method of off-line processing (see also Section 8.6.2) is used in nearly all cases published so far. However, this comes along with restrictions, such as focusing on rudimentary or special functions in VLC transmission. For example, the mitigation of background optical noise produced by ambient AC-powered LED lamps or fluorescent light sources was investigated in an experimental VLC setup using off-line processed DC-biased DMT, as reported in [76]. The results have shown that the noise produced by AC-powered LEDs has a negligible effect, as the lowest-frequency subcarrier is far above the 50 or 60 Hz effects, while the noise from fluorescent lamps can be removed by excluding the impaired subcarriers. Moreover, the mitigation of phosphor layer effects caused by the YB-LED-based Tx using channel estimation and one-tap equalization at the Rx was confirmed. Further examples of such experimental work based on off-line processing and mainly targeting high bit rates in a laboratory environment can be found in [8–10, 77–79]. However, more complex subjects including the influence of channel variations and adaptive bit rate link control are hard to investigate under real-time conditions using off-line experiments. If restrictions in DSP speed and bandwidth are accepted, a simple alternative may be to employ programmable hardware such as a DSP development kit. An example of that kind is presented in [40], where it was the aim to investigate the performance of DC-biased DMT transmission in the presence of (infrared) LED non-linearity via a hardware demonstrator including real-time implementation of the digital signal processing.

8.5   Performance enhancement of DMT modulation

In recent years, numerous approaches have been reported aiming at performance enhancement of the discussed modulation schemes, in terms of spectral efficiency and power efficiency as well as PAPR. In the following, a brief overview is given.

8.5.1   Combination of ACO-OFDM and DC-biased DMT modulation

It seems rather obvious to combine the ACO-OFDM modulation scheme, which only uses the odd subcarriers, with a complementary scheme in order to utilize the even subcarriers too. Such an approach, called *asymmetrically clipped DC-biased* optical OFDM (ADO-OFDM) is presented in [80]. In this technique, the odd subcarriers are modulated using ACO-OFDM while the even ones are modulated using DC-biased DMT. The DMT component does not cause interference on the odd frequency subcarriers. Hence, a conventional receiver can demodulate the ACO-OFDM component after detection. On the other hand, the ACO-OFDM signal causes clipping noise, which affects the even subcarriers. This interference can be estimated at the Rx, and thus cancelled at the expense of a 3 dB noise penalty in the DC-biased DMT component. Clearly, ADO-OFDM provides better bandwidth efficiency than ACO-OFDM, since all subcarriers are used to carry data. Simulation results for an AWGN channel have shown that this method can also achieve better optical power performance than the conventional OFDM schemes [72].

8.5.2   Spectrally factorized OFDM

Another means for improving the optical efficiency of OFDM for IM/DD transmission has been proposed in [81]. The formalism for non-negative multiple subcarrier modulation described there is denoted as *spectrally factorized* optical OFDM (SFO-OFDM). Instead of sending data directly on the subcarriers, the autocorrelation of the complex data sequence is performed in this scheme, which guarantees non-negativity without explicit bias. SFO-OFDM covers all band-limited OFDM signals, and unlike ACO-OFDM, it uses the entire available bandwidth for data modulation. Moreover, it mitigates the high PAPR, which is typical in PAM-DMT and ACO-OFDM systems. According to [81], 0.5 dB in optical power is gained over ACO-OFDM at a BER of 10^{-5}.

8.5.3   Flip-OFDM

A further unipolar modulation approach is known as flip-OFDM [82]. In *flip-OFDM*, the positive and negative parts are extracted from the real bipolar OFDM time domain signal, and transmitted in two consecutive OFDM symbols. Since the negative part is flipped before transmission, both subframes have positive samples, as required in IM/DD systems. The basic flip-OFDM scheme, as presented in [83], performs a compression of the time samples in order to sustain the duration of the original bipolar symbol frame. Consequently, bandwidth and data rate are doubled, while the length of the CP is reduced by 50% when compared to ACO-OFDM. As an alternative, the parameters of an ACO-OFDM system can be maintained by omitting the (not imperatively required) compression of the OFDM subframes. In that way, it has been shown by simulation that both the ACO- and flip-OFDM schemes have the same spectral efficiency and BER performance in the electrical domain. However, flip-OFDM offers savings of 50% in computational receiver complexity over the ACO-OFDM scheme, in particular for slow fading channels [82].

8.5.4   Unipolar OFDM

The so-called *unipolar* OFDM modulation scheme (U-OFDM) has been developed with the aim of reducing the PAPR and to close the 3 dB gap between OFDM and ACO-OFDM for bipolar signals, whilst generating a unipolar signal, which does not require the biasing of DC-biased DMT [84]. The modulation process of U-OFDM starts with conventional modulation of an OFDM signal. After obtaining a real bipolar signal, it is transformed into a unipolar one by encoding the absolute value of each time sample and its polarity into a pair of new time samples (absolute value on one out of two possible positions depending on the polarity). At the Rx, reconstruction of the original bipolar OFDM signal is straightforward and the process continues with conventional demodulation of an OFDM signal.

Since each time sample from the original OFDM signal is encoded into a sample pair of the U-OFDM signal, the spectral efficiency is the same as of ACO-OFDM, but U-OFDM has better power efficiency in an AWGN channel.

8.5.5   Position modulating OFDM

Based on how the symbols are assigned to the subcarriers, the unipolar operation forms unipolar optical OFDM symbols such as the various operations used to generate the ACO-OFDM, the flip-OFDM, the U-OFDM (see above), or the so-called *position modulation* OFDM (PM-OFDM) described in [85]. PM-OFDM utilizes the DFT approach but removes the Hermitian symmetry constraint. This is done at the Tx by splitting the IFFT output signal corresponding to its real and imaginary components. Their positive and negative parts are further separated, and the two negative signals are flipped by polarity inversion. The resulting four real and positive signals are then sequentially transmitted. Based on this Tx technique, two Rx structures are presented in [85] targeting either high BER performance or low Rx-complexity.

In both, LOS and diffused channels the low-complexity Rx (including time domain MMSE equalizer) achieves the same BER performance as an ACO- or flip-OFDM system, while the total system complexity is significantly lower. On the other hand, the high-performance Rx includes a frequency domain MMSE equalizer and thus an extra FFT and IFFT block, resulting in a marginally higher overall transceiver complexity compared to an ACO-OFDM system. As outlined in [85], this extra effort yields ~ 4 dB improvement at a BER of 10^{-4} in a diffuse optical wireless channel.

8.5.6   Diversity-combined OFDM

When only the odd subchannels are loaded as in ACO-OFDM modulation, the clipping distortion falls exclusively onto the even subchannels, as mentioned above. This natural separation of signal and distortion indicates that at the Rx side, some extent of spectral diversity among even and odd subchannels can be observed. Based on this fact, the idea of an *asymmetrically clipped, diversity-combined* OFDM (AC/DC-OFDM) system has been presented in [86]. There it is revealed theoretically and by simulation, that the effective SNR at the Rx of an ACO-OFDM system can be improved significantly at the expense of one additional IFFT-FFT pair and a diversity combining decoding algorithm at the Rx. As diversity combining adds two different signal components, while an additional noise cancellation would reduce the noise at the Rx, it might be expected that combining these techniques could give further performance gains. This is, however, not true as was shown analytically in [87]. On the other hand, it is important to note that noise cancellation giving a maximum improvement of 3 dB is more computationally efficient than diversity combining.

8.5.7   Further approaches

In order to enhance the overall performance of VLC systems and to efficiently exploit spectrum resources by means of MSM, various additional proposals have been published. Two of them are briefly addressed in this section.

Multicarrier code division multiple access (MC-CDMA) is a transmission scheme that combines the robustness of orthogonal modulation and the flexibility of CDMA schemes. In MC-CDMA, an individual user's complex-valued data symbol is spread over OFDM subcarriers in the frequency domain using a spreading code. The symbols from different users are aggregated in the frequency domain and passed to the OFDM modulator for transformation into the time domain. The further steps of modulation and IM/DD transmission are the same as in the case of DC-biased DMT. At the Rx side, CP removal, FFT, and de-spreading are carried out, respectively. In [88], an optimized selection of active subcarriers is proposed in order to significantly increase the power efficiency in a multi-user indoor scenario. The average transmit power reduction is accomplished by selecting the subcarriers based on pre- or post-equalization, while proper setting of a fixed DC bias ensures low system complexity.

A hybrid *multi-layer modulation* (MLM) scheme designed for optical OFDM-based IM/DD transmission is introduced in [89]. In optical systems, MLM is expected to be capable of offering a fine granularity in terms of throughput versus robustness. Both the concept of MLM and related double turbo Rx algorithms are detailed in this publication. In addition, the layer-specific optimum weights that the MLM scheme should obey are conceived. Significant gains are demonstrated by comparing the MLM-aided technique to the ACO-OFDM and DC-biased DMT schemes as a function of both the electrical and the optical energy per bit to single-sided noise power spectral density. However, the benefits are achieved at the expense of an increased transceiver complexity. As stated in [89], the approach is for further study under practical optical channel conditions.

8.6   System design and implementation aspects

8.6.1   Aspects of system design

So far, research and development on VLC using advanced LED modulation has focused on physical transmission basics, where usually a directed LOS is assumed. On the other hand, non-directed LOS links are more convenient in practical use, particularly if terminal mobility is desired. Non-directed indoor wireless IR channels have been extensively studied in the past by means of modeling, simulations, and experiments. Considering existing knowledge in this regard and similarities to VLC, such studies can provide adequate guidance for designing indoor VLC systems. However, there is only a little experience in practical terms of high-speed VLC via non-directed LOS and non-LOS links, i.e., links in diffuse scenarios.

Off-line experiments using a YB-LED and DC-biased DMT modulation in a realistic non-LOS broadcast configuration are presented e.g. in [90]. It is shown there that an area of

~18 m^2 can be covered with 100–200 Mbit/s, while the brightness level at the Rx equipped with a blue filter is about 500 lux (generated by an extra LED for lighting), and thus complies with workplace requirements. Some further results achieved for the first time with a real bidirectional system in a non-LOS configuration are presented below in Section 8.6.4.

Bidirectional links are a necessary condition for any kind of communication beyond broadcasting. Various approaches for an optical wireless indoor uplink have been discussed and the development of a quality uplink is currently one of the main research topics in this field. The most common approach, irrespective of the modulation format, is to use IR light as illustrated in Fig. 8.4. Examples of medium and high-speed systems are given e.g. in [20] and [91], respectively. However, visible light uplinks are also considered, as for instance shown in Fig. 8.5, given that a visible spot in the Rx region fits the application needs.

At the same time, bidirectional links enable dynamic adaptation to changing properties of the link, e.g. caused by ambient light or LOS blocking. The time-varying channel can easily be estimated using frequency domain channel estimation, and adaptive modulation can be applied based on the requested data rates and quality of service. The idea of OWC using DMT modulation with adaptive bit and power loading was proposed for the first time in [17] and independently thereof almost simultaneously in [18]. By this method, subcarriers are loaded with the best suitable modulation depending on the channel properties at individual frequencies, i.e., the power is distributed according to the SNR needs of each chosen modulation order or format while keeping the system constraints regarding BER and transmit power satisfied. This leads to optimal utilization of the available resources in a non-flat communication channel even in the presence of non-linear distortion caused by the transfer characteristics of common LEDs. The whole concept is analytically presented in [33, 34, 92]. Related algorithms for bit and power allocation are also addressed in [93], while experimental proofs of concept can be found in [94–96]. Further experimental studies have been performed according to [97] in order to demonstrate that the benefits of adaptively controlling the resultant QAM modulation orders also apply to YB-LED-based systems where blue filtering at the Rx is omitted.

In addition to channel adaptation, the possibility to combine DMT modulation with any *multiple access schemes* makes it an excellent preference for indoor VLC applications. Initial work directed toward multi-channel systems is published in [77] (see also [10]), where subcarrier multiplexing (SCM) is used to provide multi-user capability. The downlink capacity of an experimental setup based on off-line processing was assembled to 575 Mbit/s by wavelength division multiplexing (WDM) using the colored channels of a RGB-LED, while the YB-LED uplink of the full duplex system offered 225 Mbit/s. Pre- and post-equalization in the frequency domain were adopted to compensate for Tx and channel distortions. In addition, the transmission capacity was optimized using various QAM modulation orders. By adjusting the bandwidths and modulation formats of the sub-channels, the downlink and uplink capacities can be easily reconfigured in such a setup.

Regarding the *feeder or backhaul network* of VLC, a combination with the power line infrastructure and thus with PLC suggests itself, cf. for example [1, 2, 19, 98, 99]. From the viewpoint of DMT modulation used in both systems, this idea is of particular interest as it could simplify interfaces and interworking. Komine *et al.* proposed and analyzed an integrated system of this kind in [100] based on their earlier work and subsequently, studies addressing that subject have been performed at several places. As an example, broadcasting the DMT-QAM signal of a PLC system by directly modulating the YB-LED of a VLC Tx was presented in [101]. More recent proposals for an integration of PLC and VLC can be found e.g. in [102]. For reasons of completeness, it has to be mentioned that optical fibers and DMT transmission can also serve for VLC backhauling. However, this creates a special area of research employing e.g. laser light, and thus will not be considered further here.

The interplay of VLC, backbone network and terminal systems also calls for an appropriate *media access control* (MAC) protocol. A specific MAC layer was proposed for the optical wireless channel (considering both visible light and IR applications) in the framework of the OMEGA project [103], see also Section 8.6.4. For VLC broadcast, a simplified frame format was described. At the Tx side of the considered system, the MAC layer provided a serial data stream at 100 Mbit/s for digital signal processing and modulation on the PHY layer. After transmission via VLC link and PHY processing at the Rx side, the retrieved data stream was passed to the MAC layer. An overview of the access method, the optical wireless data link layer (OWMAC) and the MAC frame adapted to that VLC prototype is given in [104]. All the modules for MAC and PHY processing at both the Tx and Rx side were implemented in FPGA technology [104, 105]. Apart from that work, multiple access schemes for multicarrier-based VLC channels are also addressed in e.g. [106, 107].

A major design challenge regarding the commercialization of VLC for joint use in lighting systems is how to incorporate the commonly used PWM *light dimming techni-que* while maintaining reliable high-speed communication links. In [108] this problem was considered in terms of VLC power constraints due to lighting requirements in living rooms for both daytime and night-time scenarios, i.e. even in the worst case of commu-nication with the lights off. It was shown that very low light emission (virtually lights-off) is sufficient to maintain coverage at data rates up to the Mbit/s range using OOK or PPM. This leads to the conclusion that even if the average optical transmit power is severely dimmed and less power efficient modulation schemes such as DMT are used, there should be some electrical signal power for communication to a certain extent.

Regarding VLC systems based on DC-biased DMT or ACO-OFDM, dimming cannot be achieved directly, but the data signal finished for driving the LED can be combined with a dimming technique on the PHY layer. However, the simple way of multiplying the transmission signal and a PWM pulse train for dimming control during its "on" period is not an option, as the data throughput would be limited to the PWM rate, which is as low as a few 10 kHz in commercial LED lighting systems [12]. Achieving high-speed transmission with this approach is only feasible when the PWM dimming signal is at least twice the highest subcarrier frequency of the DMT signal to avoid subcarrier interference [21]. Such a constraint is hardly acceptable, as the LED modulation band-width should include the PWM frequency, which cuts the bandwidth for data trans-mission by 50%. Therefore, dimming schemes are being developed such as *reverse polarity* OFDM (RPO-OFDM), proposed in [109]. This method combines the high-frequency OFDM signal with the low-frequency PWM dimming signal, while both

Table 8.4 Key results of laboratory experiments using WDM in VLC systems based on DC-biased DMT.

Aggregate bit rate (Gbit/s)	Colors of WDM channels	Modulation bandwidth (MHz)	Number of subcarriers	Remarks	Reference
0.575	RG	50	64	PIN-Rx, 2 SCM channels per WDM channel, blue LED only used for lighting; in addition uplink via YB-LED	[10]
0.750	RGB	50	64	PIN-Rx, adaptive Nyquist windowing	[110]
0.803	RGB	50	32	APD-Rx	[111]
1.250	RGB	100	128	APD-Rx	[8]
2.930	RGB	230	471	PIN-Rx	[79]
3.400	RGB	280	512	APD-Rx	[9]

signals contribute to the effective LED brightness. RPO-OFDM utilizes the entire period of a PWM signal for data transmission by adjusting the polarity of the data symbols before they are superimposed on the PWM pulse train, which defines the dimming level, i.e. the symbol polarity is reversed during the "on" periods of the PWM signal. In this way, the data rate is maintained for a wide dimming range independently of the PWM frequency. The approach also keeps the signal within the dynamic range constraint of the LED. In conclusion, the data rate in RPO-OFDM is not limited by the PWM signal frequency, and the LED dynamic range is fully utilized to minimize the non-linear distortion of the multicarrier communication signal. This technique can be applied to any version of real-valued OFDM/DMT signal, preprocessed for transmission in VLC.

OFDM/DMT modulation continually gains in popularity due to its attractive communication performance, and the results achieved so far indicate that it is an excellent candidate for bidirectional transmission in VLC systems close to reality. Nevertheless, additional research is necessary, in particular to examine all aspects of cooperation with lighting systems and its effects on light quality in practical smart lighting scenarios.

8.6.2 DMT/OFDM application in advanced systems

When using RGB-type LEDs (or more generally, multi-color LED devices), a VLC system enables *wavelength division multiplexing* (WDM). To determine the potential of WDM in terms of transmission capacity, over recent years several laboratory studies on the WDM transmission performance in VLC systems have been carried out and reported. As a demonstration of bit rate per WDM channel was the main driver, spectrally efficient DC-biased DMT modulation was employed almost without exception.

Key results of laboratory experiments using WDM in VLC systems based on DC-biased DMT are given in Table 8.4, which indicates that bit rates of several 100 Mbit/s per WDM channel and total capacities in the Gbit/s range are possible.

It should be noted that all of these results have been achieved by exploring the WDM channels one by one and by means of off-line processing. In such a configuration, the

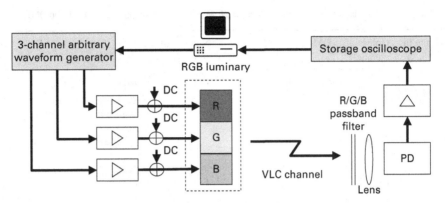

Figure 8.12   Setup for off-line evaluation of simultaneous DC-biased DMT transmission using three VLC channels in WDM fashion.

transmit signal is generated by software and an arbitrary waveform generator (AWG), while the received signal is recorded and subsequently evaluated by software. A typical setup of such an experiment is sketched in Fig. 8.12.

As VLC systems typically use multiple LEDs when combined with indoor lighting, it is an obvious step to apply the optical wireless *multiple-input multiple-output* (MIMO) principle in order to boost the overall transmission capacity. MIMO processing can compensate inter-channel crosstalk, thus allowing for parallel transmission from a number of LEDs [112]. In a corresponding optical MIMO proof of concept using AWG-generated DC-biased DMT modulation, a 2×2 Tx module consisting of four YB-LEDs, and a 3×3-channel imaging Rx equipped with a blue filter, an aggregate bit rate of 1.1 Gbit/s has been demonstrated, as reported in [113].

Optical spatial modulation (OSM) is another bandwidth and power efficient MIMO scheme for OWC. It is based on multiple spatially separated Tx units and utilizes their location to carry extra information [114], i.e., in addition to applying basic signal modulation, OSM transmits further bits in the spatial domain by considering the Tx array as an extended constellation diagram. As only one Tx is active at any given time instant, the Rx unit can easily determine the index of an active Tx and thus can be kept simple. In the case of aligned Tx and Rx units, it was shown that the paths of the optical MIMO channel are nearly uncorrelated. Thus, an accurate Tx-Rx alignment results in a power efficiency increase with respect to the modulation scheme used. A comprehensive overview of the state of the art in spatial modulation for MIMO technologies, including application in VLC, is provided e.g. by [115]. A novel unipolar modulation method for OWC based on OSM is introduced in [116]. It combines the basic spatial modulation scheme and OFDM techniques for OWC [69]. Results show that the new approach improves the power efficiency. For the same spectral efficiency, it exhibits 5 to 9 dB higher energy efficiency than ACO-OFDM, while in contrast to DC-biased DMT it eliminates the need for a DC bias. Consequently, it exhibits a considerable power efficiency gain for the same spectral efficiency.

8.6.3   Practical implementation issues

With the aim of reducing the *DSP complexity* at both the Tx and the Rx, IFFT/FFT processing can be replaced by a discrete Hartley transform (DHT) [117, 85]. DHT is a real-valued self-inverse transform without Hermitian symmetry constraint at the input. As a result, there is no need for complex algebra and the same fast DSP algorithm can be used for modulation and demodulation as well, if real constellations are used for subcarrier modulation. Real-time implementations of DHT-based Tx and Rx thus have confirmed a reduction of both complexity and computing time, while the performance is the same as in DFT-based DC-biased DMT and ACO-OFDM systems, respectively [117].

An important goal of VLC in general is to drive the LED at the best possible electro-optical power conversion efficiency. Additionally in dual use with lighting, the illumination function must be maintained with minimum extra power consumption. In most cases, VLC designs try to employ (digital) off-the-shelf *high-power LED drivers* [109], and also to use the dimming capability usually offered by such devices. An outline of the important role of LED drivers in smart lighting systems can be found in [2], while the LED driver's energy efficiency along with the OOK, VPPM and OFDM modulation schemes is discussed in [118]. Usually, the large current of high-power LEDs controlled by the driver circuit severely degrades its response. In [38] it is shown that OOK-based systems nonetheless can achieve rather sound energy efficiency while enabling transmission rates of 477 Mbit/s. In this case, the red device of a RGB-LED was driven by a specially designed circuit with pre-emphasis, which improved the 3 dB bandwidth of the optical Tx to 91 MHz. Another approach to a simple and practical LED driver providing a high overall Tx-bandwidth is based on drawing out the remaining carriers that exist in the LED depletion capacitance during the "off" state of the OOK input signal [119].

These examples for OOK modulation indicate that it appears unrealistic to use standard driver circuits in OFDM/DMT-based systems, which exploit the bandwidth more efficiently but need more complex analog circuitry, usually along with higher power consumption. Because the optical Tx performance depends, inter alia, strongly on the driver output impedance, careful impedance matching to the LED device or module is of utmost importance. As a typical example, Fig. 8.13 illustrates the bandwidth of an optical Tx including high-power LED and customized analog circuitry, which can drive currents up to 1.2 A. Such a driver was employed in various VLC setups using either a RGB-LED [8] or a YB-LED including blue filter [77, 120]. Compared to pure lighting, an increased power consumption of about 30% is observed when changing to VLC operation at the same brightness level at the Rx. In view of such power values, RF leakage can be stronger than the received optical signal. That is why accurate RF shielding of the VLC Tx in any case is an important subject.

Beyond the modulation scheme used in a VLC system, convenient ways for performance improvement are equalization at the Tx (pre-equalization), at the Rx (post-equalization), or a combination of such techniques. As an example, pre-equalization can be a part of the LED driving module [29].

Post-equalization is briefly addressed below. A noise cancellation method for an ACO-OFDM Rx is presented in [121], where the anti-symmetry of the time domain signal

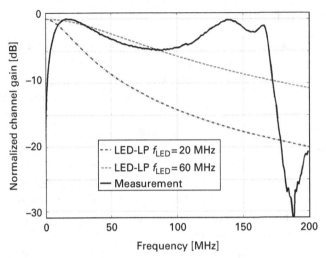

Figure 8.13 VLC channel response taken with a network analyzer at a short distance from a high-power YB-LED, which was driven by a custom-tailored analog driver circuit at a bias current of 0.7 A. The effective modulation bandwidth was about 180 MHz. For comparison simulated first order RC low-pass LED amplitude responses are also shown [120].

samples inherent in ACO-OFDM is used to identify which samples of the received signal are most likely to be due to the addition of noise. These samples are then set to zero. A maximum gain of 3 dB in optical power can be achieved with this method. The same pairwise maximum likelihood (ML) Rx structure can be employed in PAM-DMT systems by considering their anti-symmetry properties. It is worth mentioning that the use of this Rx technique in a system based on flip-OFDM results in the U-OFDM scheme as presented in [84], cf. Sections 8.5.3 and 8.5.4. In another case, post-processing is proposed to eliminate the effect of noise outside the maximum channel delay by means of a least square (LS) channel estimation method [122]. The scheme is based on a comb-type pilot subcarrier arrangement, where each OFDM symbol has pilot tones at periodically-located subcarriers. After ordinary LS channel estimation, the derived channel is truncated in the time domain according to the maximum channel delay via DFT and inverse DFT, respectively. In this way, the method can significantly improve the BER performance, as has been proved by simulation at different QAM constellation orders in a DC-biased DMT system.

Apart from the limited modulation bandwidth, white LEDs suffer from intrinsic non-linearity. This effect is particularly detrimental when high PAPR modulation formats such as DMT are employed. Much work has been done to overcome this impairment, cf. Section 8.3.3. In [123], the application of Volterra equalization is proposed for compensating ISI and the YB-LED non-linearity in a VLC system that employs M-PAM modulation. It was demonstrated that a Rx using a decision feedback equalizer (DFE) with non-linear Volterra feed-forward section up to the second order can efficiently compensate effects of non-linearity of the transmitting LED and performs up to 5 dB better in terms of optical power than using a standard DFE. In view of such results, this scheme is also attractive for DMT-based VLC systems.

It is hardly surprising that the SNR distribution in an area covered by indoor VLC can be improved, as well as the spectral efficiency of the DMT-based modulation, by tilting the (moving) receiver plane. Such a scheme has been proposed in [124] to enhance the performance in an adaptive system, where a feedback channel is used to adapt bit and power loading. The optimum tilting angle of the photodetector is determined by the Newton method, which is a fast algorithm requiring a three-step search from the initial state.

Up to now, experimental comparisons of different system approaches and details thereof are extremely scarce. Quite recently, the bit rate performances of various VLC approaches, including DC-biased DMT, ACO-OFDM, and U-OFDM have been experimentally compared in [125]. Clearly, the bandwidth efficiency of DC-biased DMT presents a higher bit rate performance, but is of course less power efficient than ACO-OFDM. On the other hand, the ACO-OFDM as well as the U-OFDM schemes suffer from the effect of baseline wander in practical transmission, caused by the moving average of the asymmetric signal. Such findings require closer examination in future work.

8.6.4   Implementation and demonstration

The feasibility of white LED intensity modulation using OFDM was validated for the first time by experimental results as published in [126]. Since then, a lot of research and analytical work, as well as simulations and experimental studies on DMT/OFDM-based VLC, including subsystems thereof as discussed above, have been carried out. Despite this, there are only a very small number of system implementations which incorporate real-time signal processing for DMT/OFDM modulation, etc. This is however mandatory for system verification under real-life conditions and an important step for paving the way to real commercial application.

An early proof-of-concept hardware demonstrator is shown in Fig. 8.14, where the bandwidth of 45 kHz was limited by the DSP development kit used for performing OFDM-related signal processing. The aim of this system was to study the performance of phase-incoherent optical OFDM under various conditions, e.g. against different electrical SNRs. Moreover, the system served for characterization of the wireless channel and for modeling.

In the course of the European research and development project OMEGA, a fully-fledged high-speed VLC system based on DMT modulation has been developed and implemented, which offers wireless broadcast channels of 100 Mbit/s (net) in homes [129]. Under real-life conditions such a performance was demonstrated for the first time in February 2011, by broadcasting up to four HD video streams (80% utilization of the 100 Mbit/s channel) simultaneously from a total of 16 YB-LED ceiling lamps to a photodetector placed anywhere within the lit area of about 10 m². An illustration is given in Fig. 8.15. The MAC protocol (Optical Wireless MAC) developed especially for this purpose, and digital signal processing functionalities for the PHY layer, were implemented on FPGA boards. The PHY processing encompassed a typical DMT modulator/demodulator including scrambler and forward error correction (FEC)

Table 8.5 Digital PHY parameters of the fully-fledged VLC demonstrator.

Parameter	Value
Signal bandwidth (MHz)	30.5947
Number of subcarriers (incl. DC)	32
Length of cyclic prefix (samples)	4
QAM order	16
FEC (Reed Solomon)	187 207

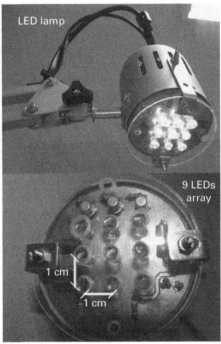

Figure 8.14 *Left*: Laboratory VLC demonstrator for initial unidirectional OFDM-based real-time experiments [127]. *Right*: Reading lamp with an array of nine white LEDs in order to enlarge the coverage area for OFDM experiments [128].

encoder/decoder with parameters summarized in Table 8.5. The physical layer also featured full synchronization on bit and frame levels [104, 130, 131].

More recently at Fraunhofer HHI, a bidirectional high-speed real-time VLC system has been presented (Fig. 8.16), which operates in a time division duplex mode. Rate-adaptive DC-biased DMT is implemented by feedback via the reverse link. The transceivers are equipped with proprietary VLC Tx and Rx modules providing a modulation bandwidth of up to 180 MHz. However, the bandwidth used is below 100 MHz, limited by the DSP chips. These transceiver modules can operate without active cooling and offer 1000BASE-T Ethernet interfaces, cf. Fig. 8.16 (bottom).

Figure 8.15 The first public demonstration of a high-speed VLC system using DMT modulation for broadcasting multiple video streams at a data rate of 100 Mbit/s (125 Mbit/s gross data rate at PHY layer). (*Top*) Array of VLC transmitters on the ceiling; (*bottom*) mobile receiver device consisting of a photodetector and an amplifier for forwarding the signal to the demodulator and further to the video screens.

A particular advantage of this real-time VLC system is the variable throughput of up to 500 Mbit/s with controlled BER, depending on the quality of the optical communication channel. The system offered the first mobile VLC experience. As shown in Fig. 8.17, the most important parameter is the light intensity at the Rx, leading to an almost proportional data rate adaptation. It should be noted that the observed

Figure 8.16   (*Top*): Two bidirectional transceivers communicating via a diffuse link. Different colors are used for down and uplink as an example, but also the same colors can be used. (*Bottom left*): View of the diffusing spot of the link at a (standard painted) wall at a distance of up to ~3 m from the transceivers. The overall link length is about 6 m. (*Bottom right*): New generation of bidirectional transceiver with a form factor of 87×114×42 mm^3 (without lenses).

performance is similar with illumination LEDs of any color. IR-LEDs can be used as well as white light phosphorous LEDs, which benefit from built-in pre-equalization and thus do not require blue filtering at the Rx. The power consumption of this system is only moderately increased – by 30% compared to the original lighting function – due to an optimized LED driver design.

Figure 8.17 reveals another important result: *high-speed non-LOS data transmission*. In this particular case, light was reflected by the white-painted wall to the Rx. More than 100 Mbit/s are possible at a link range of about 3 m, despite diffusion through reflection.

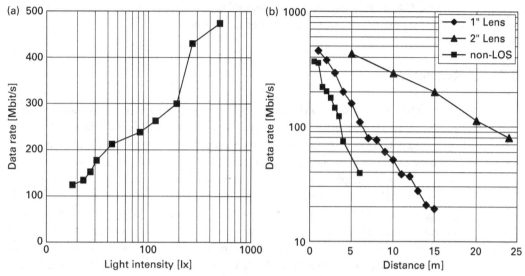

Figure 8.17   (a) Maximum achieved data rates with the real-time rate-adaptive VLC system depending on the light intensity at the Rx. (b) Maximum data rates achieved with different link types: LOS link with narrow FOV (triangle curve), LOS link with wide FOV (diamond curve), and a non-LOS diffuse link (square curve).

In this way, the feasibility of a flexible and robust VLC connection was demonstrated at high bit rates under different link scenarios.

A similar implementation was published recently in [91]. This VLC system used a YB-LED device for the downlink and an IR uplink, with an integrated OFDM-based DSP chip for real-time operation. The LED bandwidth of ~1 MHz has been increased to ~12 MHz by an analog pre-equalization technique without using a blue filter. The modulation bandwidth of the DSP chip applied is between 2 and 30 MHz and the adaptive OFDM modulation uses formats from QPSK to 16-QAM. This system offers a data throughput up to 37 Mbit/s over about 1.5 m distance.

From all demonstrations discussed above it can be concluded that rate-adaptive systems based on DMT modulation, as also used in radio and wired transmission, can be implemented with off-the-shelf components. It has been shown that such systems are an excellent solution for robust optical wireless-link systems with peak data rates up to 500 Mbit/s under various lighting conditions. Nonetheless, the commercially available DSP chips, which of course are not intended for application in VLC systems, or solutions extensively tailored on FPGA, respectively, limit the data rate and the overall system performance. Moreover, there are limits in optimization of important parameters such as power consumption. Full custom integrated circuits using current technology would be able to remove such shortcomings but would require entering the mass market.

These experiments and measurements using real-time adaptive bidirectional VLC systems fulfill an important intermediate step to turn the optical WiFi vision into reality.

The next steps will integrate VLC technology into real-life room lighting, and will implement extensions such as Mp-to-Mp functionality.

8.7   Summary

In this chapter, the use of DMT modulation schemes in VLC for indoor application is presented along with the most important features and conditions. Items of VLC are discussed with focus on the provision of high-speed links using DMT-based modulation, i.e. only as far as there are special requirements of system design in that respect. In accordance with the present status of research and development on OWC around the world, the PHY layer is emphasized in this chapter.

Against this backdrop, the optical wireless channel is briefly characterized, and methods to exploit its capacity by means of typical high-brightness LEDs intended for lighting in combination with DMT-based modulation and IM/DD transmission are presented. Properties of the basic DMT/OFDM versions, namely DC-biased DMT, ACO-OFDM and PAM-DMT are discussed and compared. Moreover, several variants of these schemes for use in VLC systems are highlighted. In a nutshell, DMT/OFDM modulation is identified as a method of choice for exploiting VLC channels of restricted bandwidth due to the LED device and the driver characteristics. The modulation schemes are able to take advantage of the high SNR that is available in VLC scenarios based on room lighting and typical brightness levels. A further important advantage is the ability to adapt the modulation format of each subcarrier to the SNR of the channel, commonly known as bit and power loading. Such an advanced exploitation of the channel capacity also provides flexibility in coverage, but it requires a feedback link.

In addition, it has been shown that DMT modulation can also be applied to most advanced VLC transmission schemes using WDM and MIMO techniques. By means of WDM the transmission capacity can be multiplied, however, multi-color LEDs as well as more complex or colored Rx units are necessary in such a scenario. Higher bit rates are also possible using MIMO technologies. Related research is in an early phase but at a high level of activity.

The numerous comparative analyses of the DMT/OFDM modulation formats show that both of the basic schemes using asymmetrical clipping, i.e., ACO-OFDM and PAM-DMT, are best at low spectral efficiency and high power efficiency, i.e. where power consumption rather than bit rate is of importance. Their computational complexity is nearly the same. However, these schemes as well as variants proposed and under discussion have not yet left the laboratory stage due to the difficulty of real-time implementation, which is mandatory for VLC verification under conditions such as mobility. On the other hand, DC-biased DMT has the lowest computational complexity among the three major DMT/OFDM modulation formats. The scheme is expected to offer the highest data rate in applications, where the additional DC bias power required in VLC to create a non-negative modulation signal can provide a complementary functionality, such as illumination. In fact, the record-breaking results in transmission rate as presented in [8, 9, 11, 46] have all been achieved using DC-biased DMT. The first implementations and transceiver prototypes, as far as they have been made public, also use DC-biased DMT including real-time signal processing, thus enabling a mobile VLC experience. According to this, the suitability of DMT modulation for robust indoor VLC application at high data rates has been confirmed already by real-life verification and public demonstration.

It is also important to note that the poor modulation bandwidth of white light LEDs of the YB-type is not an issue any more. It has been demonstrated that the blue filter, which mitigates the adverse effect of the phosphor layer but is unfavorable concerning the power budget, can be replaced by equalization. In that way, the optical power reaching the Rx can be fully exploited at bit rates of several 100 Mbit/s or even more.

Although DMT-based modulation has continually gained in popularity due to its attractive performance, various versions of single-carrier frequency domain equalization (SCFDE) transmission move into the VLC spotlight. These techniques offer reduced PAPR, which may result in an overall better performance [132, 133]. Newly, DC-biased DMT and SCFDE transmission have been compared in a WDM experimental setup, where aggregate data rates of 2.5 Gbit/s and 3.75 Gbit/s respectively, have been achieved [134]. Hence, the SCFDE scheme clearly outperformed DMT in that setup. Another single-carrier modulation scheme, namely carrier-less amplitude and phase (CAP) modulation has been explored too [79]. An aggregate bit rate of 3.22 Gbit/s, again achieved in a WDM experiment, confirms the significance of single-carrier modulation as an important option for high-speed VLC systems. Ways for further performance improvement on PHY level may also be opened by a quasi balanced detection scheme for OFDM signals, recently introduced in [135].

So far work on both VLC and LED-based lighting has focused on vital issues of each individual domain. More specifically, lifetime, color, luminous efficacy, etc. of devices have been subjects investigated in LED lighting development, while topics such as LED modulation and driving, related signal processing and link control have been addressed in communications. True VLC system design, however, needs multi-disciplinary work covering communications and lighting technology as well. For example, LED long-term behavior under VLC operation including a potentially high PAPR needs to be verified, as so far there is only initial experience. Regarding LED modulation and power supplies, it may be useful to drive LED lamps as an integral part of the existing infrastructure directly with AC-power. Basic results on such a VLC approach, yet at low speed and using OOK, have been published recently in [136]. Additional research is also necessary to examine LED modulation effects on light quality in practical smart lighting scenarios. As is correctly pointed out in [12] for the dual-use case, VLC and light intensity control are in conflict. Thus, it is important to consider DMT-compatible (hybrid) dimming schemes in extended VLC standards. They should also take into account emerging standards on radio and wired transmission, which are closely related from the VLC systems point of view. It is an open issue whether lighting and energy demands will be met in all respects. Thus, this presents a field for further studies. More verification in real environments is also necessary, which calls for joint work by the communications and lighting industries.

Several VLC techniques have been verified by experiments with reliable off-line processing and also by a few demonstrations using components off the shelf. Real low-cost systems now need custom designed devices, such as integrated DSP for optimization of VLC performance and features to be considered, as stated in this chapter. In this matter, interaction with DSP chip manufacturers is necessary along with ongoing work of standardization at the system level.

Finally, it must be mentioned that system integration has been considered only at a rudimentary level so far and mobility-related issues such as handover in VLC have not been adequately addressed. Moreover, a dedicated standardization roadmap is essential for future availability of VLC in portable devices [137]. Standardization activities so far emanate from the Infrared Data Association (IrDA) interest group and from the IEEE. Whereas the IrDA mainly provides specifications for wireless infrared protocols, the IEEE has published a first OWC standard, IEEE 802.15.7–2011, for VLC. The recent extension of the International Telecommunication Union (ITU) g.hn standard (ITU-T Recommendation G.9960, 2011) foreseeing an optical channel is also of importance, cf. [102], concerning integration of VLC, e.g. into home networks and cooperation with present infrastructure.

Even if there are still challenges as named above, which are being faced by research and development, VLC presents a realistic and promising supplementary technology to radio communication.

References

[1] M. Kavehrad, "Sustainable energy-efficient wireless applications using light," *IEEE Communications Magazine*, **48**, (*12*), 66–73, 2010.

[2] A. Sevincer, A. Bhattarai, M. Bilgi, M. Yuksel, and N. Pala, "LIGHTNETs: Smart lighting and mobile optical wireless networks – a survey," *IEEE Communications Surveys & Tutorials*, **15**, (*4*), 1620–1641, 2013.

[3] S. Haruyama, "Visible light communication using sustainable LED lights," Proceedings of 5th ITU Kaleidoscope: *Building Sustainable Communities*, 2013.

[4] L. Hanzo, H. Haas, S. Imre, *et al.*, "Wireless myths, realities, and futures: From 3G/4G to optical and quantum wireless," *Proceedings of the IEEE*, **100**, 1853–1888, 2012.

[5] D.K. Borah, A.C. Boucouvalas, C.C. Davis, S. Hranilovic, and K. Yiannopoulos, "A review of communication-oriented optical wireless systems," *EURASIP Journal on Wireless Communications and Networking*, **91**, 1–28, 2012.

[6] K.-D. Langer, J. Hilt, D. Schulz, *et al.*, "Rate-adaptive visible light communication at 500 Mb/s arrives at plug and play," *Optoelectronics and Communications* SPIE Newsroom, DOI 10.1117/2.1201311.005196, 2013.

[7] J. Vučić, C. Kottke, S. Nerreter, *et al.*, "230 Mbit/s via a wireless visible-light link based on OOK modulation of phosphorescent white LEDs," *OFC/NFOEC Technical Digest* 2010, paper OThH3.

[8] C. Kottke, J. Hilt, K. Habel, J. Vučić, and K.-D. Langer, "1.25 Gbit/s visible light WDM link based on DMT modulation of a single RGB LED luminary," Proc. European Conference and Exhibition on *Optical Communication (ECOC)* 2012, paper We.3.B.4.

[9] G. Cossu, A.M. Khalid, P. Choudhury, R. Corsini, and E. Ciaramella, "3.4 Gbit/s visible optical wireless transmission based on RGB LED," *Optics Express*, **20**, (*26*), B501–B506, 2012.

[10] Y. Wang, Y. Shao, H. Shang, *et al.*, "875-Mb/s asynchronous bi-directional 64QAM-OFDM SCM-WDM transmission over RGB-LED-based visible light communication system," *OFC/ NFOEC Technical Digest* 2013, paper OTh1G.3.

[11] D. Tsonev, H. Chun, S. Rajbhandari, *et al.*, "A 3-Gb/s single-LED OFDM-based wireless VLC link using a Gallium Nitride μLED," *IEEE Photonics Technology Letters*, **26**, (*7*), 637–640, 2014.

[12] J. Gancarz, H. Elgala, and T.D.C. Little, "Impact of lighting requirements on VLC systems," *IEEE Communications Magazine*, **51**, (*12*), 34–41, 2013.

[13] R.D. Roberts, S. Rajagopal, and S.-K. Lim, "IEEE 802.15.7 physical layer summary," Proc. IEEE GLOBECOM Workshops 2011, pp. 772–776.

[14] S. Rajagopal, R.D. Roberts, and S.-K. Lim, "IEEE 802.15.7 visible light communication: Modulation schemes and dimming support," *IEEE Comm. Mag.*, **50**, (*3*), 72–82, 2012.

[15] D. Tsonev, S. Videv, and H. Haas, "Light fidelity (Li-Fi): Towards all-optical networking," Proc. SPIE **9007**, *Broadband Access Communication Technologies VIII*, 900702, 2013.

[16] J.M. Kahn and J.R. Barry, "Wireless infrared communications," *Proceedings of the IEEE*, **85**, (*2*), 265–298, 1997.

[17] J. Grubor, V. Jungnickel, K.-D. Langer, and C. v. Helmolt, "Dynamic data-rate adaptive signal processing method in a wireless infra-red data transfer system," Patent EP1897252 B1, 24 June 2005.

[18] O. Gonzalez, R. Perez-Jimenez, S. Rodriguez, J. Rabadan, and A. Ayala, "OFDM over indoor wireless optical channel," *IEE Proc. Optoelectronics*, **152**, (*4*), 199–204, 2005.

[19] L. Grobe, A. Paraskevopoulos, J. Hilt, *et al.*, "High-speed visible light communication systems," *IEEE Communications Magazine*, **51**, (*12*), 60–66, 2013.

[20] Z. Ghassemlooy, H. Le Minh, P.A. Haigh, and A. Burton, "Development of visible light communications: Emerging technology and integration aspects," Proc. *Optics and Photonics Taiwan International Conference (OPTIC)*, 2012.

[21] G. Ntogari, T. Kamalakis, J.W. Walewski, and T. Sphicopoulos, "Combining illumination dimming based on pulse-width modulation with visible-light communications based on discrete multitone," *Journal of Optical Comm. and Networking*, **3**, (*1*), 56–65, 2011.

[22] J. Grubor, S.C.J. Lee, K.-D. Langer, T. Koonen, and J. Walewski, "Wireless high-speed data transmission with phosphorescent white-light LEDs," Proc. European Conference and Exhibition on *Optical Communication (ECOC)* 2007, **6**, Post-Deadline Paper PD3.6.

[23] C.W. Chow, C.H. Yeh, Y.F. Liu, and Y. Liu, "Improved modulation speed of LED visible light communication system integrated to main electricity network," *El. Letters*, **47**, (*15*), 2011.

[24] Y. Pei, S. Zhu, H. Yang, *et al.*, "LED modulation characteristics in a visible-light communication system," *Optics and Photonics Journal*, **3** (*2B*), 139–142, 2013.

[25] J. Grubor, S. Randel, K.-D. Langer, and J.W. Walewski, "Broadband information broadcasting using LED-based interior lighting," *J. of Lightwave Tech.*, **26**, (*24*), 3883–3892, 2008.

[26] J. Armstrong, R.J. Green, and M.D. Higgins, "Comparison of three receiver designs for optical wireless communications using white LEDs," *IEEE Communications Letters*, **16**, (*5*), 748–751, 2012.

[27] D.C. O'Brien, L. Zeng, H. Le-Minh, *et al.*, "Visible light communications," in R. Kraemer, M.D. Katz (eds.), *Short-Range Wireless Communications*, pp. 329–342, Wiley & Sons Ltd., 2009.

[28] C.W. Chow, C.H. Yeh, Y. Liu, and Y.F. Liu, "Digital signal processing for light emitting diode based visible light communication," *IEEE Phot. Society Newsletter*, **26**, (*5*), 9–13, 2012.

[29] H. Le-Minh, D.C. O'Brien, G. Faulkner, *et al.*, "80 Mbit/s visible light communications using pre-equalized white LED," Proc. 34th European Conference and Exhibition on *Optical Communication (ECOC)*, 2008.

[30] J. Vučić, C. Kottke, S. Nerreter, K.-D. Langer, and J.W. Walewski, "513 Mbit/s visible light communications link based on DMT-modulation of a white LED," *Journal of Lightwave Technology*, **28**, (*24*), 3512–3518, 2010.

[31] H. Chun, C.-J. Chiang, and D.C. O'Brien, "Visible light communication using OLEDs: Illumination and channel modeling," Int. Workshop on *Optical Wireless Communications (IWOW)*, 2012.

[32] P.A. Haigh, Z. Ghassemlooy, I. Papakonstantinou, and H. Le Minh, "2.7 Mb/s with a 93 kHz white organic light emitting diode and real time ANN equalizer," *IEEE Photonics Technology Letters*, **25**, (*17*), 1687–1690, 2013.

[33] J. Grubor and K.-D. Langer, "Efficient signal processing in OFDM-based indoor optical wireless links," *Journal of Networks*, **5**, (*2*), 197–211, 2010.

[34] J. Vučić, "Adaptive modulation technique for broadband communication in indoor optical wireless systems," PhD Thesis at Technische Universitaet Berlin, Germany, 2009.

[35] X. Li, J. Vučić, V. Jungnickel, and J. Armstrong, "On the capacity of intensity-modulated direct-detection systems and the information rate of ACO-OFDM for indoor optical wireless applications," *IEEE Transactions on Communications*, **60**, (*3*), 799–809, 2012.

[36] S. Dimitrov and H. Haas, "Information rate of OFDM-based optical wireless communication systems with nonlinear distortion," *J. of Lightwave Tech.*, **31**, (*6*), 918–929, 2013.

[37] X. Zhang, K. Cui, M. Yao, H. Zhang, and Z. Xu, "Experimental characterization of indoor visible light communication channels," Proc. 8th International Symposium on *Communication Systems, Networks & Digital Signal Processing (CSNDSP)*, 2012.

[38] N. Fujimoto and H. Mochizuki, "477 Mbit/s visible light transmission based on OOK-NRZ modulation using a single commercially available visible LED and a practical LED driver with a pre-emphasis circuit," *OFC/NFOEC Technical Digest* 2013, paper JTh2A.73.

[39] I. Neokosmidis, T. Kamalakis, J.W. Walewski, B. Inan, and T. Sphicopoulos, "Impact of nonlinear LED transfer function on discrete multitone modulation: Analytical approach," *Journal of Optical Communications and Networking*, **1**, (*5*), 439–451, 2009.

[40] I. Stefan, H. Elgala, R. Mesleh, D. O'Brien, and H. Haas, "Optical wireless OFDM system on FPGA: Study of LED nonlinearity effects," Proc. 73rd IEEE *Vehicular Technology* Conference (VTC Spring), pp. 1–5, 2011.

[41] S.-B. Ryu, J.-H. Choi, J. Bok, H. Lee, and H.-G. Ryu, "High power efficiency and low nonlinear distortion for wireless visible light communication," Proc. 4th IFIP International Conference on *New Technologies, Mobility and Security (NTMS)*, pp. 1–5, 2011.

[42] D. Lee, K. Choi, K.-D. Kim, and Y. Park, "Visible light wireless communications based on predistorted OFDM," *Optics Communications*, **285**, (*7*), 1767–1770, 2012.

[43] D. Tsonev, S. Sinanovic, and H. Haas, "Complete modeling of nonlinear distortion in OFDM-based optical wireless communication," *Journal of Lightwave Technology*, **31**, (*18*), 3064–3076, 2013.

[44] R. Mesleh, H. Elgala, and H. Haas, "LED nonlinearity mitigation techniques in optical wireless OFDM communication systems," *Journal of Optical Communications and Networking*, **4**, (*11*), 865–875 , 2012.

[45] R. Mesleh, H. Elgala, and H. Haas, "On the performance of different OFDM based optical wireless communication systems," *Journal of Optical Communications and Networking*, **3**, (*8*), 620–628, 2011.

[46] A.M. Khalid, G. Cossu, R. Corsini, P. Choudhury, and E. Ciaramella, "1-Gb/s transmission over a phosphorescent white LED by using rate-adaptive discrete multitone modulation," *IEEE Photonics Journal*, **4**, (*5*), 1465–1473, 2012.

[47] B. Inan, S.C.J. Lee, S. Randel, *et al.*, "Impact of LED nonlinearity on discrete multitone modulation," *Journal of Optical Communications and Networking*, **1**, (*5*), 439–451, 2009.

[48] C. Ma, H. Zhanga, K. Cuib, M. Yaoa, and Z. Xu, "Effects of LED lighting degradation and junction temperature variation on the performance of visible light communication," International Conference on *Systems and Informatics (ICSAI)*, pp. 1596–1600, 2012.

[49] J.B. Carruthers and J.M. Kahn, "Multiple-subcarrier modulation for non-directed wireless infrared communication," *IEEE J. on Selected Areas in Comm.*, **14**, (*3*), 538–546, 1996.

[50] Y. Tanaka, T. Komine, S. Haruyama, and M. Nakagawa, "A basic study of optical OFDM system for indoor visible communication utilizing plural white LEDs as lighting," 8th Int. Symp. on *Microwave and Optical Technol. (ISMOT)*, pp. 303–306, 2001.

[51] J. Armstrong and A.J. Lowery, "Power efficient optical OFDM," *Electronics Letters*, **42**, (*6*), 370–372, 2006.

[52] S.C.J. Lee, S. Randel, F. Breyer, and A.M.J. Koonen, "PAM-DMT for intensity-modulated and direct-detection optical communication systems," *IEEE Photonics Technology Letters*, **21**, (*23*), 1749–1751, 2009.

[53] S. Randel, F. Breyer, and S.C.J. Lee, "High-speed transmission over multimode optical fibers," Proc. 34th European Conference and Exhibition on *Optical Communication (ECOC)* 2008, paper OWR2.

[54] J.M. Cioffi, "A multicarrier primer," *ANSI Contribution T1E1*, **4**, 91–157, 1991.

[55] J. Armstrong, "OFDM for optical communications," *Journal of Lightwave Technology*, **27**, (*3*), 189–204, 2009.

[56] A.V. Oppenheim and R.W. Schafer, *Discrete-Time Signal Processing*, Prentice-Hall, 1989.

[57] S.K. Hashemi, Z. Ghassemlooy, L. Chao, and D. Benhaddou, "Orthogonal frequency division multiplexing for indoor optical wireless communications using visible light LEDs," Proc. 6th Int. Symp. on *Communication Systems, Networks & Digital Signal Processing (CSNDSP)* 2008, pp. 174–178.

[58] J. Armstrong, B.J.C. Schmidt, D. Kalra, H.A. Suraweera, and A.J. Lowery, "Performance of asymmetrically clipped optical OFDM in AWGN for an intensity modulated direct detection system," Proc. IEEE *Global Telecommunications* Conf. (GLOBECOM '06), SPC07–4, 2006.

[59] S.C.J. Lee, F. Breyer, S. Randel, H.P.A. van den Boom, and A.M.J. Koonen, "High-speed transmission over multimode fiber using discrete multitone modulation," *Journal of Optical Networking*, **7**, (*2*), 183–196, 2008.

[60] E. Vanin, "Signal restoration in intensity-modulated optical OFDM access systems," *Optics Letters*, **36**, (*22*), 4338–4340, 2011.

[61] X. Li, R. Mardling, and J. Armstrong, "Channel capacity of IM/DD optical communication systems and of ACO-OFDM," Proc. Int. Conf. on *Communications* (ICC) 2007, pp. 2128–2133, 2007.

[62] S.K. Wilson and J. Armstrong, "Transmitter and receiver methods for improving asymmetrically-clipped optical OFDM," *IEEE Trans. on Wireless Comm.*, **8**, (*9*), 4561–4567, 2009.

[63] S.C.J. Lee, F. Breyer, S. Randel, *et al.*, "Discrete multitone modulation for maximizing transmission rate in step-index plastic optical fibers," *Journal of Lightwave Technology*, **27**, (*11*), 1503–1513, 2009.

[64] L. Chen, B. Krongold, and J. Evans, "Performance analysis for optical OFDM transmission in short-range IM/DD systems," *Journal of Lightwave Technology*, **30**, (*7*), 974–983, 2012.

[65] S. Tian, K. Panta, H.A. Suraweera, *et al.*, "A novel timing synchronization method for ACO-OFDM-based optical wireless communications," *IEEE Transactions on Wireless Communications*, **7**, (*12*), 4958–4967, 2008.

[66] M.M. Freda and J.M. Murray, "Low-complexity blind timing synchronization for ACO-OFDM-based optical wireless communications," Proc. IEEE GLOBECOM Workshops 2010, pp. 1031–1036.

[67] R. You and J.M. Kahn, "Average power reduction techniques for multiple subcarrier intensity-modulated optical signals," *IEEE Trans. Communications*, **49**, (*12*), 2164–2171, 2001.

[68] B. Ranjha and M. Kavehrad, "Precoding techniques for PAPR reduction in asymmetrically clipped OFDM based optical wireless system," Proc. SPIE **8645**, *Broadband Access Communication Technologies VII*, **86450**R, 2013.

[69] J. Armstrong and B.J.C. Schmidt, "Comparison of asymmetrically clipped optical OFDM and DC-biased optical OFDM in AWGN," *IEEE Comm. Letters*, **12**, (*5*), 343–345, 2008.

[70] D.J.F. Barros, S.K. Wilson, and J.M. Kahn, "Comparison of orthogonal frequency-division multiplexing and pulse-amplitude modulation in indoor optical wireless links," *IEEE Transactions on Communications*, **60**, (*1*), 153–163, 2012.

[71] S. Dimitrov, S. Sinanovic, and H. Haas, "Signal shaping and modulation for optical wireless communication," *Journal of Lightwave Technology*, **30**, (*9*), 1319–1328, 2012.

[72] S.D. Dissanayake and J. Armstrong, "Comparison of ACO-OFDM, DCO-OFDM and ADO-OFDM in IM/DD systems," *Journal of Lightwave Technology.*, **31**, (*7*), 1063–1072, 2013.

[73] Z. Yu, R.J. Baxley, and G.T. Zhou, "EVM and achievable data rate analysis of clipped OFDM signals in visible light communication," *EURASIP Journal on Wireless Communications and Networking*, (*1*), 1–16, 2012.

[74] L. Chen, B. Krongold, and J. Evans, "Theoretical characterization of nonlinear clipping effects in IM/DD optical OFDM systems," *IEEE Transactions on Communications*, **60**, (*8*), 2304–2312, 2012.

[75] S. Dimitrov, S. Sinanovic, and H. Haas, "Clipping noise in OFDM-based optical wireless communication systems," *IEEE Trans. on Communications*, **60**, (*4*), 1072–1081, 2012.

[76] C.W. Chow, C.H. Yeh, Y.F. Liu, and P.Y. Huang, "Background optical noises circumvention in LED optical wireless systems using OFDM," *IEEE Phot. J.*, **5**, (*2*), 7900709, 2013.

[77] C. Kottke, K. Habel, L. Grobe, *et al.*, "Single-channel wireless transmission at 806 Mbit/s using a white-light LED and a PIN-based receiver," Proc. 14th Int. Conf. on *Transparent Optical Networks* (ICTON), paper We.B4.1, 2012.

[78] Y. Wang, Y. Wang, N. Chi, J. Yu, and H. Shang, "Demonstration of 575-Mb/s downlink and 225-Mb/s uplink bi-directional SCM-WDM visible light communication using RGB LED and phosphor-based LED," *Optics Express*, **21**, (*1*), 1203–1208, 2013.

[79] F.M. Wu, C.T. Lin, C.C. Wei, *et al.*, "Performance comparison of OFDM signal and CAP signal over high capacity RGB-LED-based WDM visible light communication," *IEEE Photonics Journal*, **5**, (*4*), 7901507, 2013.

[80] S.D. Dissanayake, K. Panta, and J. Armstrong, "A novel technique to simultaneously transmit ACO-OFDM and DCO-OFDM in IM/DD systems," Proc. IEEE GLOBECOM Workshops 2011, pp. 782–786.

[81] K. Asadzadeh, A.A. Farid, and S. Hranilovic, "Spectrally factorized optical OFDM," Proc. 12th Canadian Workshop on *Information Theory (CWIT)*, pp. 102–105, 2011.

[82] N. Fernando, Y. Hong, and E. Viterbo, "Flip-OFDM for optical wireless communications," Proc. *Information Theory Workshop (ITW)*, 5–9, 2011.

[83] Y.-I. Jun, "Modulation and demodulation apparatuses and methods for wired / wireless communication," Patent WO/2007/064165, 2007.

[84] D. Tsonev, S. Sinanovic, and H. Haas, "Novel unipolar orthogonal frequency division multiplexing (U-OFDM) for optical wireless," IEEE 75th *Vehicular Technology* Conference (VTC Spring), pp. 1–5, 2012.

[85] A. Nuwanpriya, A. Grant, S.-W. Ho, and L. Luo, "Position modulating OFDM for optical wireless communications," Proc. 3rd IEEE Workshop on *Optical Wireless Communications (OWC'12)*, pp. 1219–1223, 2012.

[86] L. Chen, B. Krongold, and J. Evans, "Diversity combining for asymmetrically clipped optical OFDM in IM/DD channels," Proc. IEEE *Global Telecomm. Conf.* (GLOBECOM '09), pp. 1–6, 2009.

[87] S.D. Dissanayake, J. Armstrong, and S. Hranilovic, "Performance analysis of noise cancellation in a diversity combined ACO-OFDM system," Proc. 14th Int. Conf. on *Transparent Optical Networks (ICTON)*, 2012.

[88] M.Z. Farooqui and P. Saengudomlert, "Transmit power reduction through subcarrier selection for MC-CDMA-based indoor optical wireless communications with IM/DD," *EURASIP Journal on Wireless Communications and Networking*, (*1*), 1–14, 2013.

[89] R. Zhang and L. Hanzo, "Multi-layer modulation for intensity modulated direct detection optical OFDM," *J. of Optical Communications and Networking*, **5**, (*12*), 1402–1412, 2013.

[90] G. Cossu, A.M. Khalid, R. Corsini, and E. Ciaramella, "Non-directed line-of-sight visible light system," *OFC/NFOEC Technical Digest* 2013, paper OTh1G.2.

[91] C.H. Yeh, Y.-L. Liu, and C.-W. Chow, "Real-time white-light phosphor-LED visible light communication (VLC) with compact size," *Opt. Express*, **21**, (*22*), 26192–26197, 2013.

[92] J. Grubor, V. Jungnickel, and K.-D. Langer, "Adaptive optical wireless OFDM system with controlled asymmetric clipping," IEEE Proc. 41st Asilomar Conference on *Signals, Systems and Computers*, 2007.

[93] Z. Sun, Y. Zhu, and Y. Zhang, "The DMT-based bit-power allocation algorithms in the visible light communication," Proc. 2nd International Conference on *Business Computing and Global Informatization*, pp. 572–575, 2012.

[94] K.-D. Langer, J. Vučić, C. Kottke, *et al.*, "Advances and prospects in high-speed information broadcast," Proc. 11th Int. Conf. on *Transparent Optical Networks (ICTON)*, paper Mo.B5.3, 2009.

[95] J. Vučić, C. Kottke, S. Nerreter, *et al.*, "White light wireless transmission at 200+ Mb/s net data rate by use of discrete-multitone modulation," *IEEE Photonics Technology Letters*, **21**, (*20*), 1511–1513, 2009.

[96] D. Bykhovsky and S. Arnon, "An experimental comparison of different bit-and-power-allocation algorithms for DCO-OFDM," *Journal of Lightwave Technology*, **32**, (*8*), 1559–1564, 2014.

[97] C.W. Chow, C.H. Yeh, Y.F. Liu, P.Y. Huang, and Y. Liu, "Adaptive scheme for maintaining the performance of the in-home white-LED visible light wireless communications using OFDM," *Optics Communications*, **292**, (*1*), 49–52, 2013.

[98] K.L. Sterckx, "Implementation of continuous VLC modulation schemes on commercial LED spotlights," Proc. 9th International Conference on *Electrical Engineering/Electronics, Computer, Telecommunications and Information Technology (ECTI-CON)*, 2012.

[99] H. Elgala, R. Mesleh, and H. Haas, "Indoor optical wireless communication: Potential and state-of-the-art," *IEEE Communications Magazine*, **49**, (*9*), 56–62, 2011.

[100] T. Komine, S. Haruyama, and M. Nakagawa, "Performance evaluation of narrowband OFDM on integrated system of power line communication and visible light wireless communication," Proc. 1st Int. Symp. on *Wireless Pervasive Computing*, 2006.

[101] S.E. Alavi, A.S.M. Supa'at, S.M. Idrus, and S.K. Yusof, "New integrated system of visible free space optic with PLC," Proc. 3rd Workshop on *Power Line Communications (WSPLC)*, 2009.

[102] H. Ma, L. Lampe, and S. Hranilovic, "Integration of indoor visible light and power line communication systems," Proc. 17th IEEE International Symposium on *Power Line Communications and its Applications (ISPLC)*, pp. 291–296, 2013.

[103] K.-D. Langer, J. Grubor, O. Bouchet, *et al.*, "Optical wireless communications for broadband access in home area networks," Proc. 10th Int. Conf. on *Transparent Optical Networks (ICTON)*, **4**, 149–154, 2008.

[104] O. Bouchet, P. Porcon, and E. Gueutier, "Broadcast of four HD videos with LED ceiling lighting: Optical-wireless MAC," Proc. SPIE **8162**, *Free-Space and Atmospheric Laser Communications XI*, 81620L, 2011.

[105] O. Bouchet, P. Porcon, J.W. Walewski, *et al.*, "Wireless optical network for a home network," Proc. SPIE **7814**, *Free-Space Laser Communications X*, 781406, 2010.

[106] M.V. Bhalerao, S.S. Sonavane, and V. Kumar, "A survey of wireless communication using visible light," *Int. Journal of Advances in Engineering & Technology*, **5**, (*2*), 188–197, 2013.

[107] J. Dang and Z. Zhang, "Comparison of optical OFDM-IDMA and optical OFDMA for uplink visible light communications," Proc. International Conference on *Wireless Communications & Signal Processing (WCSP)*, 2012.

[108] T. Borogovac, M.B. Rahaim, M. Tuganbayeva, and T.D.C. Little, "Lights-off visible light communications," Proc. IEEE GLOBECOM Workshops 2011, pp. 797–801.

[109] H. Elgala and T.D.C. Little, "Reverse polarity optical-OFDM (RPO-OFDM): Dimming compatible OFDM for gigabit VLC links," *Optics Express*, **21**, (*20*), 24288–24299, 2013.

[110] R. Li, Y. Wang, C. Tang, *et al.*, "Improving performance of 750-Mb/s visible light communication system using adaptive Nyquist windowing," *Chinese Optics Letters*, **11**, (*8*), 080605/1–4, 2013.

[111] J. Vučić, C. Kottke, K. Habel, and K.-D. Langer, "803 Mbit/s visible light WDM link based on DMT modulation of a single RGB LED luminary," *OFC/NFOEC Technical Digest* 2011, paper OWB6.

[112] A.H. Azhar, T.-A. Tran, and D. O'Brien, "Demonstration of high-speed data transmission using MIMO-OFDM visible light communications," Proc. IEEE GLOBECOM Workshops 2010, pp. 1052–1056.

[113] A.H. Azhar, T.-A. Tran, and D. O'Brien, "A gigabit/s indoor wireless transmission using MIMO-OFDM visible-light communications," *IEEE Photonics Tech. Letters*, **25**, 171–174, 2013.

[114] X. Zhang, S. Dimitrov, S. Sinanovic, and H. Haas, "Optimal power allocation in spatial modulation OFDM for visible light communications," Proc. IEEE 75th *Vehicular Technology* Conference (VTC Spring), pp. 1–5, 2012.

[115] X. Di Renzo, H. Haas, A. Ghrayeb, S. Sugiura, and L. Hanzo, "Spatial modulation for generalized MIMO: Challenges, opportunities, and implementation," *Proceedings of the IEEE*, **102**, (*1*), 56–103, 2014.

[116] Y. Li, D. Tsonev, and H. Haas, "Non-DC-biased OFDM with optical spatial modulation," Proc. IEEE 24th Int. Symp. on *Personal, Indoor and Mobile Radio Communications (PIMRC)*, pp. 486–490, 2013.

[117] M.S. Moreolo, R. Muñoz, and G. Junyent, "Novel power efficient optical OFDM based on Hartley transform for intensity-modulated direct-detection systems," *Journal of Lightwave Technology*, **28**, (*5*), 798–805, 2010.

[118] G. del Campo Jiménez and F.J. López Hernándeza, "VLC oriented energy efficient driver techniques," Proc. SPIE **8550**, *Optical Systems Design* 2012, 85502F.

[119] T. Kishi, H. Tanaka, Y. Umeda, and O. Takyu, "A high-speed LED driver that sweeps out the remaining carriers for visible light communications," *Journal of Lightwave Technology*, **32**, (*2*), 239–249, 2014.

[120] L. Grobe and K.-D. Langer, "Block-based PAM with frequency domain equalization in visible light communications," Proc. IEEE GLOBECOM Workshops 2013, pp. 1075–1080, 2013.

[121] K. Asadzadeh, A. Dabbo, and S. Hranilovic, "Receiver design for asymmetrically clipped optical OFDM," Proc. IEEE GLOBECOM Workshops 2011, pp. 777–781.

[122] X. Yang, Z. Min, T. Xiongyan, W. Jian, and H. Dahai, "A post-processing channel estimation method for DCO-OFDM visible light communication," Proc. 8th Int. Symp. on *Communication Systems, Networks & Digital Signal Processing (CSNDSP)*, 2012.

[123] G. Stepniak, J. Siuzdak, and P. Zwierko, "Compensation of a VLC phosphorescent white LED nonlinearity by means of Volterra DFE," *IEEE Photonics Technology Letters*, **25**, (*16*), 1597–1600, 2013.

[124] Z. Wang, C. Yu, W.-D. Zhong, and J. Chen, "Performance improvement by tilting receiver plane in M-QAM OFDM visible light communications," *Optics Express*, **19**, (*14*), 13418–13427, 2011.

[125] A.H. Azhar and D. O'Brien, "Experimental comparisons of optical OFDM approaches in visible light communications," Proc. IEEE GLOBECOM Workshops 2013, pp. 1076–1080, 2013.

[126] M.Z. Afgani, H. Haas, H. Elgala, and D. Knipp, "Visible light communication using OFDM," Proc. 2nd International Conference on *Testbeds and Research Infrastructures for the Development of Networks and Communities (TRIDENTCOM)*, pp. 129–134, 2006.

[127] H. Elgala, R. Mesleh, H. Haas, and B. Pricope, "OFDM visible light wireless communication based on white LEDs," Proc. 65th IEEE *Vehicular Technol.* Conf. (VTC Spring), pp. 2185–2189, 2007.

[128] H. Elgala, R. Mesleh, and H. Haas, "Indoor broadcasting via white LEDs and OFDM," *IEEE Transactions on Consumer Electronics*, **55**, (*3*), 1127–1134, 2009.

[129] J. Vučić, L. Fernández, C. Kottke, K. Habel, and K.-D. Langer, "Implementation of a real-time DMT-based 100 Mbit/s visible-light link," Proc. European Conference and Exhibition on *Optical Communication (ECOC)* 2010, paper We.7.B.1.

[130] K.-D. Langer, J. Vučić, C. Kottke, *et al.*, "Exploring the potentials of optical-wireless communication using white LEDs," Proc. 13th Int. Conf. on *Transparent Optical Networks (ICTON)*, paper Tu.D5.2, 2011.

[131] J. Vučić and K.-D. Langer, "High-speed visible light communications: State-of-the-art," *OFC/NFOEC Technical Digest* 2012, paper OTh3G.3.

[132] M. Wolf, L. Grobe, M.R. Rieche, A. Koher, and J. Vučić, "Block transmission with linear frequency domain equalization for dispersive optical channels with direct detection," Proc. 12th Int. Conf. on *Transparent Optical Networks (ICTON)*, paper Th.A3.4, 2010.

[133] K. Acolatse, Y. Bar-Ness, and S.K. Wilson, "Novel techniques of single-carrier frequency-domain equalization for optical wireless communications," *EURASIP Journal on Advances in Signal Processing*, 2011, article ID 393768.

[134] N. Chi, Y. Wang, Y. Wang, X. Huang, and X. Lu, "Ultra-high-speed single red-green-blue light-emitting diode-based visible light communication system utilizing advanced modulation formats," *Chinese Optics Letters*, **12**, (*1*), 010605/1–4, 2014.

[135] Y. Wang, N. Chi, Y. Wang, *et al.*, "High-speed quasi-balanced detection OFDM in visible light communication," *Optics Express*, **21**, (*23*), 27558–27564, 2013.

[136] Y.F. Liu, C.H. Yeh, C. W. Chow, and Y.L. Liu, "AC-based phosphor LED visible light communication by utilizing novel signal modulation," *Optical and Quantum Electronics*, **45**, (*10*), 1057–1061, 2013.

[137] G. Dede, T. Kamalakis, and D. Varoutas, "Evaluation of optical wireless technologies in home networking: an analytical hierarchy process approach," *Journal of Optical Communications and Networking*, **3**, (*11*), 850–859, 2011.

9 Image sensor based visible light communication

Shinichiro Haruyama and Takaya Yamazato

9.1 Overview

Image sensors are used in digital cameras and a large number of imaging devices for industrial, media, medical, and consumer electronics applications. An image sensor consists of a number of pixels; each pixel contains a photodiode (PD), which is generally used as a receiver in visible light communications (VLC). Thus, an image sensor consisting of a number of pixels can also be used as a VLC receiver. A particular advantage of using image sensors, due to the massive number of available pixels, is the ability to spatially separate sources. Owing to the spatial separation of multiple sources, the VLC receiver uses only the pixels that sense LED transmission sources, discarding other pixels, including those sensing ambient noise. The ability to spatially separate sources also provides an additional feature to VLC, i.e., the ability to receive and process multiple transmitting sources.

In this chapter, we introduce VLC using an image sensor [1]. After presenting an overview of image sensors in Section 9.2, we introduce the use of an image sensor as a VLC receiver in Section 9.3. We provide a design of an image sensor based VLC system in Section 9.4. In Sections 9.5 and 9.6, we respectively introduce the following two unique applications of VLC systems using an image sensor: (1) massively parallel visible light transmission that can theoretically achieve a maximum data rate of 1.28 Gigabits per second; and (2) accurate sensor pose estimation that cannot be realized by a VLC system using a single-element PD. Applications of image sensor based communication are also presented in Section 9.7 for traffic signal communication, position measurements in civil engineering, and bridge position monitoring. Finally, in Section 9.8, we summarize our conclusions.

9.2 Image sensors

An image sensor is a device that converts an optical image into an electronic signal. Image sensors are used in digital cameras, camera modules, video recorders, and other imaging devices. As we describe later, image sensors can also be used as VLC receivers.

An image sensor consists of $n \times m$ pixels, ranging from 320×240 (QVGA) to $157\,000 \times 18\,000$ (line scanner). Each pixel contains a photodetector and devices for readout circuits. The pixel size ranges from 3×3 to $15 \times 15\ \mu m^2$, limited by the dynamic range and cost of the optics. The fraction of pixel area occupied by the photodetector is called the fill factor. The fill factor ranges from 0.2 to 0.9; a high fill factor is desirable. The readout circuits included in pixel devices determine the sensor conversion gain, which is the output voltage per photon collected by the PD. The readout speed determines the frame rate, which is typically 30 frames per second (fps). High frame rates are required for many industrial and measurement applications. Needless to say, VLC must use a high-frame-rate image sensor.

Two major types of image sensors are the charged coupled device (CCD) image sensor and the complementary metal oxide semiconductor (CMOS) image sensor [2,3]. Aside from CCD and CMOS sensors, a two-dimensional PD array is often used for massively parallel VLC.

9.2.1   CCD image sensor

Figure 9.1 shows the block diagram of a CCD image sensor. In the figure, when exposure is complete, a CCD sequentially transfers each pixel's charge packet. Then, the charge is converted into voltage and directed off-chip. In CCDs, incident photons are converted into electric charges and accumulated during the exposure time in a photodetector. Because CCDs use optimized photodetectors, they can offer high uniformity, low

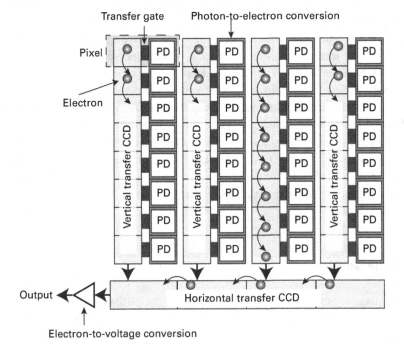

Figure 9.1   A CCD image sensor.

noise levels, and low dark current in combination with a high fill factor, high quantum efficiency, and high sensitivity; however, they do suffer from a number of drawbacks. For example, they cannot be integrated with other analog or digital circuits, such as clock generators, control circuits, or analog-to-digital convertors. Moreover, they require high amounts of power, and due to the required increase in transfer speeds, their frame rate is limited, especially for large sensors.

9.2.2 CMOS image sensor

Figure 9.2 shows the block diagram of a CMOS image sensor. These sensors have gained popularity in recent years because of their advances in multi-functionalization, low manufacturing costs, and low power consumption. The key element of a CMOS image sensor is the PD, which is one component of a pixel. PDs are typically organized in an orthogonal grid. During operation, light (photons) passing through a lens strikes the PD, where it is converted into a voltage signal and passed through an analog-to-digital converter. The converter output is often referred to as a luminance. Since a CMOS image sensor is composed of a PD array, PD outputs, i.e., light intensity or luminance values, are arranged in a square matrix to form a digital electronic representation of the scene.

The primary difference between CCD and CMOS image sensors is the readout architecture; for CCDs, charge is shifted out by the vertical and the horizontal transfer CCD, while for CMOS image sensors, charge or voltage is read out using a row and column decoder, similar to a digital memory.

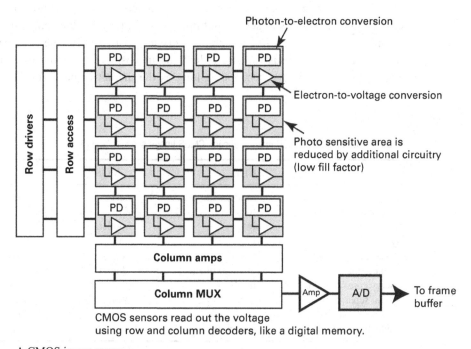

Figure 9.2 A CMOS image sensor.

9.2.3   Comparing CCD image sensors, CMOS image sensors, and photodiodes (PD)

Table 9.1 compares CCD image sensors, CMOS image sensors, and PDs. For communication, since PD is much faster than the CCD and CMOS image sensors, a single PD is attractive and has been a frequently used receiver. The PD is easy to fabricate and has low production costs; however, it usually has a non-linear response to light. For transmissions exceeding Gbps speeds, a two-dimensional PD array is used for massively parallel VLC.

Advantages of CCDs are the resulting high-quality imaging, i.e., CCDs have optimized photodetectors, very low noise, and no fixed pattern noise; however, CCDs cannot integrate with other analog or digital circuits, including clock generators, controllers, or analog-to-digital converters. High power consumption and limited frame rates, especially for large sensors, are also disadvantages of CCDs.

The speed of the readout processes for CMOS image sensors could be increased by selecting a specific set of pixels, while discarding others. Sampling a small number of relevant pixels dramatically increases readout speeds. Using this technique or other related techniques, frame rates of 10 000 fps or more are possible [5].

The potential for product integration is another advantage of CMOS image sensors over CCDs. This potential makes possible the realization of a complete single-chip camera with timing logic, exposure control, and analog-to-digital conversion. Furthermore, integration of communication pixels with conventional image pixels has also become possible [4]. Of late, most chips found in computers and other electric goods are manufactured using CMOS technology. Chip manufacturing plants cost millions of dollars to set up, but provided that they produce chips in sufficient quantities, the resultant price per chip is very low, particularly when compared with other technologies. Thus, drastic reductions in costs are possible even for high-frame-rate CMOS image sensors, as long as the market demand for such chips is high.

Table 9.1 Comparison of CCD image sensors, CMOS image sensors, and PD

	CCD	CMOS	PD
Speed	moderate to fast	fast	very fast
Sensitivity	high	low	high
Noise	low	moderate	low
System complexity	high	low	very low
Sensor complexity	high	low	very low
Chip output	analog voltage	digital bits	analog voltage
Energy consumption	moderate	low	low
Spatial separation	yes	yes	no
Product integration	low	high	no
Production cost	moderate	very low	very low

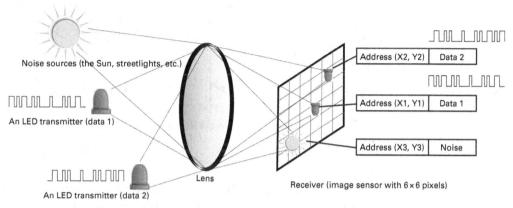

Address (X2, Y2)   Data 2

Address (X1, Y1)   Data 1

Address (X3, Y3)   Noise

Noise sources (the Sun, streetlights, etc.)

An LED transmitter (data 1)

An LED transmitter (data 2)

Lens

Receiver (image sensor with 6 × 6 pixels)

Figure 9.3   Advantages of the image sensor based VLC.

9.3   Image sensor as a VLC receiver

A CMOS image sensor can be used as a VLC receiver [1,4]. A particular advantage of using a CMOS image sensor, due to the massive number of available pixels, is its ability to spatially separate sources. Here sources include both noise sources, such as the Sun, streetlights, and other ambient lights, and transmission sources, i.e., LEDs.

The ability to spatially separate sources also provides an additional feature to VLC, i.e., the ability to receive and process multiple transmitting sources. As shown in Fig. 9.3, the data transmitted from two different LED transmitters can be captured simultaneously. Further, if a source is composed of multiple LEDs, parallel data transmission can be accomplished by independently modulating each LED.

The output of the CMOS image sensor forms a digital electronic representation of the scene, which provides unique opportunities that cannot be realized by a single-element PD or radio-wave technology. For example, it offers the ability to simultaneously utilize a multitude of image and video processing techniques, such as position estimation, object detection, and moving target detection, via the data reception capability of a VLC link.

As another example, consider a situation in which a receiver is equipped with a CMOS image sensor. To begin with, a VLC signal is captured along with its spatial position (X,Y) or the actual row and column position of a pixel. This means that the VLC signal can be represented not only by a time domain signal, but also by the direction of the incoming vector from the transmitter to the receiver. Consequently, requisite position data, which can be obtained via GPS or another position estimation system, will be available inherently through VLC transmission.

In the following subsections, we present important technical features associated with an image sensor used as a VLC receiver.

9.3.1   Temporal sampling

The Nyquist–Shannon sampling theorem, sometimes referred to as the Shannon–Someya sampling theorem, is a fundamental theorem that applies to time-dependent

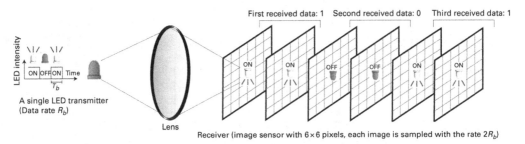

Figure 9.4   A single LED transmitter and image sensor receiver.

signals. This theorem states that if an LED transmitter generates a discrete-time signal with the time interval T_s, i.e., the data rate $R_s = 1/T_s$ sample-per-second (bps), then the frame rate of an image sensor must be greater than or equal to $2R_s$ fps. If the frame rate is less than $2R_s$ fps, then temporal aliasing occurs and the signals become indistinguishable, resulting in a situation in which the original signal is difficult to reconstruct.

Figure 9.4 shows an example of an LED transmitter with on-off-keying (OOK) modulation and an image sensor receiver. The transmitter consists of multiple LEDs generating the same signal, which can be used in this case.

Suppose that the frame rate of a CMOS image sensor is 30 fps, then the transmission rate, or equivalently the blinking rate, of an LED has to be ≤15 Hz to meet the sampling theorem. Only at blinking rates ≤15 Hz, can the receiver recognize the blinking.

If the frame rate is 1000 fps, then the LED blinking rate becomes 500 Hz. Blinking at such a fast rate is invisible to the human eye, thus we recognize LED sources as continuous illuminating devices that do not blink. The use of high-frame-rate image sensors is mandatory for VLCs.

9.3.2   Spatial sampling

The ability to spatially separate sources also provides an additional feature to VLCs, i.e., the ability to receive and process multiple transmitting sources. As shown in Fig. 9.3, data transmitted from multiple LEDs (data 1 and data 2) can be simultaneously captured; this allows parallel data transmission by separately modulating each LED of a source comprising multiple LEDs [4].

Figure 9.5 shows an example of a 3 × 3 LED array transmitter modulated in OOK format with each LED transmitting a different signal. This is one of the advantages of using an image sensor as a VLC reception device: parallel data transmission can be achieved by separately modulating each LED.

The sampling theorem described above is for one-dimensional discrete-time signals, but it can be easily extended to images using real numbers to represent the relative intensities of pixels (picture elements).

Similarly to one-dimensional discrete-time signals, images also may suffer from aliasing if the sampling resolution or pixel density is inadequate. For example, if the distance between LEDs is too small, aliasing can occur. In general, four pixels for each

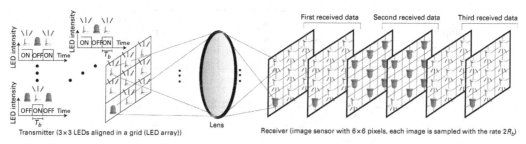

Figure 9.5   A 3 × 3 LED array transmitter and an image sensor receiver.

LED (i.e., two each for the row and the column) are required. The resultant aliasing appears as a Moiré pattern or loss in higher spatial frequency components of the image. The solution for better sampling in the spatial domain in this case would be to move the receiver closer to the transmitter, use a higher-resolution image sensor, or use a telescopic lens to enlarge the transmitter's image. In Subsection 9.4.5 below, we discuss the effect of communication distance and spatial sampling.

9.3.3   Maximal achievable data rate

In this subsection, we evaluate the maximal achievable data rate ϑ of image sensor based VLC. To begin with, consider an image sensor with $N \times M$ pixels, each pixel producing G-level gray scale signals. Let F_r be the frame rate. Then the maximal achievable data rate is obtained as $\vartheta = \frac{1}{8} NM \log_2 G \, F_r$, with the 1/8 factor as the rate reduction by the sampling of three-dimensional signals.

For example, if we consider a QVGA (320 × 240 pixels) 256 gray scale image sensor with a frame rate of 1000 fps, then the maximal achievable data rate reaches 76.8 Mbps using a 160 × 120 LED array transmitter. The latest Photoron FASTCAM SA-X2 captures 12-bit gray scale images at a frame rate of 720 000 fps with an image size of 256 × 8 pixels [5]. In this case, the data rate reaches 2.2 Gbps using a 128 × 4 LED array transmitter.

Note that the data rate obtained above is for the image sensor designed for digital imaging in which pixels are densely organized. If each pixel is placed with a gap to avoid spatial aliasing and the reception speed is much faster than the transmission data rate to avoid temporal aliasing (as is the case in Section 9.5 below with massively parallel visible light transmission), then we may drop the 1/8 factor and the rate is governed by the transmitter. For example, for an $N \times M$ LED array transmitter with each LED producing a G-level luminance signal with data rate R_b, the rate becomes $\vartheta = NM \log_2 G \, R_b$.

9.4   Design of an image sensor based VLC system

9.4.1   Transmitter

In the design of an image sensor based VLC system, we may use a single or a multiple LED transmitter (i.e., an LED array, as shown in Fig. 9.6). Here, we focus on the latter, i.e., the LED array transmitter.

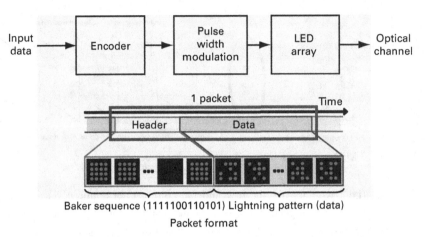

Figure 9.6   Basic structure of a multiple LED transmitter (i.e., an LED array).

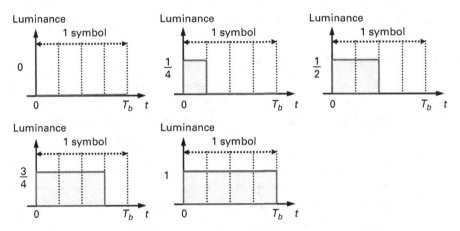

Figure 9.7   Pulse width modulation (PWM) to assign different LED luminance values.

Consider an LED array transmitter that consists of $M \times N$ LEDs. The transmitter generates a non-negative binary pulse with width T_b, where T_b is the bit duration. The data rate will be $R_b = (1/T_b)$. Hence, since each LED transmits a different bit, the bit rate of the transmitter becomes MNR_b. As shown in the figure, the luminance of the LED can be modulated by changing the width of T_b or via the pulse width modulation (PWM) technique [6]. For example, the PWM technique shown in Fig. 9.7 produces a five-level set of luminance values, i.e., 0, 1/4, 1/2, 3/4, and 1 (maximum luminance). Omitting the 0 luminance signal to avoid an entirely dark period, a four-level signal is produced. Accordingly, the PWM produces G-level gray scale signals and the data rate becomes $MN \log_2 G R_b$. Finally, the PWM signal is converted into a two-dimensional signal, and each LED transmits data in parallel by individually modulating its luminance. In other words, we transmit data as a two-dimensional LED pattern.

Note that the packet format is shown below the transmitter in Fig. 9.6. A unique code, such as the Baker code sequence, may be used for temporal synchronization.

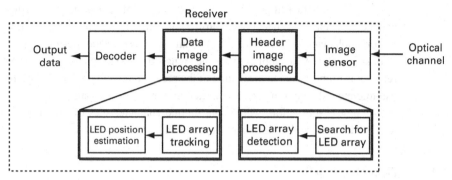

Figure 9.8   Receiver for image sensor based VLC.

9.4.2   Receiver

As shown in Fig. 9.8, the receiver consists of an image sensor, an image processing unit, and a data detection unit. The image size of the image sensor must be larger than the transmitter LED array so as to avoid spatial aliasing. Further, the frame rate must be at least twice as fast as the data rate or blinking rate of each LED.

The transmitted signal arrives at the image sensor receiver through the optical channel. After the header image-processing unit, the signal is fed to the data image-processing unit, which consists of an LED array tracking unit, an LED position estimation unit, and a luminance extraction unit. Simple template matching can be used for the tracking.

After LED array tracking, LED position estimation is performed, locating the position of each of the LEDs based on the pixel row and column, as well as the luminance values. This is only possible if accurate LED array detection is achieved, because both the shape of the LED array and the LED array tracking are necessary for the output. Finally, the output of the data image-processing unit is fed to a decoder, which outputs the retrieved data.

9.4.3   Channel

It is generally accepted that VLC links depend on the existence of an uninterrupted line-of-sight (LOS) path between the transmitter and receiver. In contrast, radio links are typically susceptible to large fluctuations in received signal amplitudes and phases. Unlike radio-waves, VLC does not suffer from multipath fading, which significantly simplifies the design of VLC links. Because VLC signals travel in a straight line between the transmitter and receiver, they can be blocked easily by vehicles, walls, or other opaque barriers. This signal confinement makes it easy to limit transmissions to receivers in close proximity. Thus, VLC networks can potentially achieve remarkably high aggregate capacity and a simplified design, because transmissions outside the communication range need not be coordinated. In other words, it is not necessary to consider sources outside the visual range.

However, VLC has several potential drawbacks. First, since visible light cannot penetrate walls or buildings, VLC coverage is restricted to small areas and some applications will therefore require the installation of access points that must be inter-connected via a wired backbone. Further, in addition to outright physical blocks, thick fog or smoke can blur visible light links and decrease system performance.

In short-range VLC applications, the signal-to-noise ratio (SNR) of a direct-detection receiver is proportional to the square of the received optical power. Therefore, VLC links can tolerate only a comparatively limited amount of signal path loss.

In the next subsection, we present unique characteristics of the image sensor based VLC channel from the viewpoint of the image sensor itself.

9.4.4   Field-of-view (FOV)

The field-of-view (FOV) is an important parameter for defining the image-capturing range of the receiver. There are three FOV types, namely horizontal FOV, vertical FOV, and diagonal FOV. We can calculate the FOV from the focal length of the lens and the image sensor size. As an example, Fig. 9.9 shows the calculation of the diagonal FOV using focal length and sensor size. The diagonal FOV is the widest of the three FOVs; thus, we define the diagonal FOV as the maximum FOV (FOV_{max}) [7].

The image-capturing range of the channel differs depending on the width of the FOV. For narrow FOVs, the receiver is equipped with a telescopic lens that captures a magnified image of a target source, usually located far away. In other words, the receiver captures the target as if the transmitter is placed in front of the receiver. Thus, the receiver can easily recognize each LED having a transmitter. In addition, the receiver is less physically affected by ambient light noise, because light other than the target transmitter has difficulty penetrating the lens in the narrow FOV. At first glance, these characteristics seem favorable; however, the number of transmitters that the receiver simultaneously recognizes decreases, because the receiver limits the number of transmitters used for communication.

In the case of a wide FOV, a large view can be captured. Unlike a narrow FOV, a wide FOV allows the receiver to simultaneously capture numerous transmitters without limiting the number of target sources. If these target transmitters send data using visible

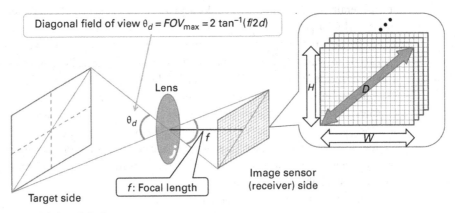

Diagonal field of view $\theta_d = FOV_{max} = 2\tan^{-1}(f/2d)$

Lens

θ_d

f

H D

W

Target side

f: Focal length

Image sensor (receiver) side

Figure 9.9   Field-of-view (FOV).

light, the receiver can obtain these data. More specifically, if the receiver can distinguish visible light from transmitters within the FOV, all data from transmitters can be received; however, the size of each target on the image is different and depends on the communication distance. The size of the target on the image decreases as the distance between transmitter and receiver increases. In addition, a receiver with a wide FOV is susceptible to more ambient light noise as compared to a receiver with a narrow FOV, because the probability that the receiver recognizes light sources other than VLC transmitters is high.

9.4.5   Effect of communication distance and spatial frequency

As described in Subsection 9.3.2 above, the size of a target (i.e., the transmitter LED array) for a captured image differs according to the communication distance. Generally, this size increases as communication distance of the target decreases. Figure 9.10 shows the LED array used to evaluate the relationship between communication distance and the number of pixels of the LED array. Let the actual distance between neighboring LEDs be d_a. We use an LED array with $d_a = 20$ mm and an image sensor receiver with a focal length of 35 mm and a resolution of 128×128. Here, we define pixel distance d_p as the distance between two neighboring LEDs on the image sensor. From Fig. 9.10, the numbers of pixels of the LED array are observed to differ depending on the communication distance. We focus on the number of LEDs on the array and the number of pixels in the image. As discussed above, to distinguish each LED on the array for an image, the number of pixels should be twice the number of LEDs [8].

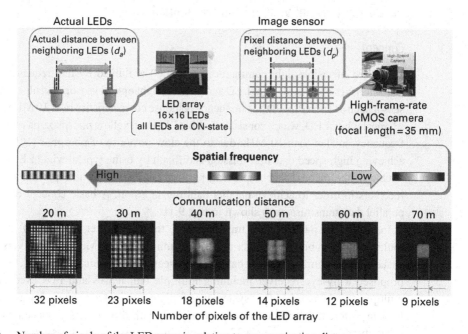

Figure 9.10   Number of pixels of the LED array in relation to communication distance.

The LED array has 256 LEDs arranged in a 16 × 16 square matrix. In this layout, the allocation of 32 × 32 pixels is necessary to distinguish each LED on the array. This requirement is satisfied when the communication distance is 20 m, as illustrated in Fig. 9.10. Next, we need to determine the number of LEDs required to distinguish each LED for various communication distances. When the communication distance is 40 m, the number of pixels is 18 × 18. In this situation, if the transmitter has LEDs arranged in a 9 × 9 or smaller square matrix, then the receiver can distinguish each LED on the image. This requirement is equivalent to sending data using 8 × 8 LEDs or every other LED of the 16 × 16 LED array. In the same manner, when the communication distance is 70 m, the requirement is 16 LEDs arranged in a 4 × 4 square matrix or less, which is equivalent to sending data using 4 × 4 LEDs, or every fourth LED of a 16 × 16 LED array.

Next, we again focus on the effect of communication distance from the viewpoint of spatial frequency. Spatial frequency refers to the number of pairs of bars imaged within the image sensor. The spatial frequency may be found by applying the two-dimensional Fourier transform. Higher resolutions result in the improved detection of high-spatial-frequency components of fine-textured objects.

As shown in Fig. 9.10, the 20 m LED array image (32 × 32 pixels) has more high-frequency components than the 70 m LED array image (9 × 9 pixels). The reduction in the number of pixels of the LED array results in the reduction of high-frequency components. In other words, the longer the distance, the more high-frequency components are lost. This reflects the fact that the channel is a low-pass channel with a cutoff frequency that decreases as the distance between transmitter and receiver increases [9].

9.5   Massively parallel visible light transmission

9.5.1   Concept

The speed of visible light communication (VLC) is limited by the frequency response characteristics of a visible light LED and a visible light optical sensor. A visible light optical sensor such as an Si PIN PD has frequency response characteristics over 10 MHz; however, a typical white LED, which consists of a blue LED and yellow phosphor, has a 3 dB cutoff frequency smaller than 10 MHz due to the slow response of phosphor [10]. Therefore, achieving high-speed data transmission is difficult by using typical white LEDs.

One of the methods for improving performance is to use multiple transmitters and receivers in order to achieve parallel data transmission. A basic concept of massively parallel data transmission is shown in Fig. 9.11.

In massively parallel data transmissions, the transmitter consists of an array of multiple LEDs, and the receiver consists of multiple PDs. Multiple receivers for free-space optical communication have been researched previously. In [11], the authors proposed a fly-eye receiver including multiple ball lenses and PDs used to receive multiple data in parallel. This system was fairly complicated because it was required to look in different directions simultaneously. In [12], the authors proposed an imaging diversity receiver employing a 37-pixel imaging receiver. This system was designed to

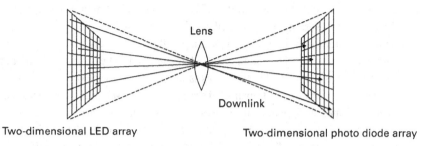

Figure 9.11   Massively parallel data transmission of visible light communication.

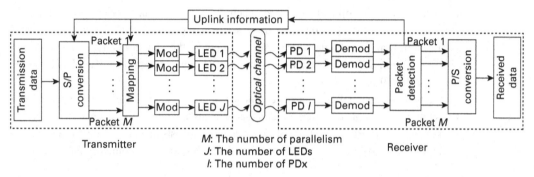

Figure 9.12   System architecture of a massively parallel data transmission system.

be used not only for LOS communication, but also non-LOS communication by selecting and combining signals from multiple PDs; however, it did not consider parallel data reception. Conversely, in both [13] and [14], the authors proposed massively parallel data transmission via a PD array consisting of 256 PDs. We describe the details of such an approach in the next section.

Conventional image sensors have a two-dimensional array of PDs; however, they usually have slow frame rates, in the range 10–30 fps, due to their mechanism of sequentially reading data from PDs, i.e., pixels. This frame rate is not fast enough for the massively parallel data transmission to achieve the desired data rates that exceed Gbps; however, there is a special type of two-dimensional PD array having each signal from individual PDs connected to a package pin such that all PD reception data can be read out in parallel. One example of such a two-dimensional PD array with parallel read-out is shown in the middle picture of Fig. 9.15.

9.5.2   System architecture

Figure 9.12 shows the system architecture of a massively parallel data transmission system. A data stream is divided into multiple packets and transmitted independently from each LED in the transmitter LED array. The receiver, which is composed of a lens and a two-dimensional image sensor, demodulates all the signals in each pixel in parallel. Haruyama's group at Keio University in Japan developed a high-speed massively parallel data transmission system using a white LED array and a two-dimensional PD array [13,14]. The details of this system are described in this subsection.

The system architecture consists of a parallel visible light transmitter, a parallel visible light receiver, and an uplink infrared connection. The link between transmitter and receiver must first be established, as is described in the next subsection. After the link is established, data to be transmitted are converted to M parallel data packets. The address that indicates the position in the two-dimensional LED array is attached to each data packet, along with an error detecting code of cyclic redundancy check (CRC). These parallel packets are allocated to J LEDs and transmitted by these LEDs. At the parallel visible light receiver, each PD of the PD array independently detects the incoming light signal, and then performs error detection to check if the received bits are correctly recovered from the optical signal for each PD. If some have errors, an Automatic Repeat reQuest (ARQ) is performed by sending the request via the uplink infrared connection. If all parallel data are correctly received, they are converted to serial data.

The configuration of the system developed by Haruyama's group at Keio University is shown in Fig. 9.13.

The two-dimensional LED array, composed of 24 × 24 white LEDs, can be used as a transmitter. The visible light receiver consists of a lens and a two-dimensional PD array that consists of 16 × 16 PIN PDs. The uplink connection consists of an infrared LED attached to the two-dimensional receiver and an infrared PD attached to the two-dimensional transmitter.

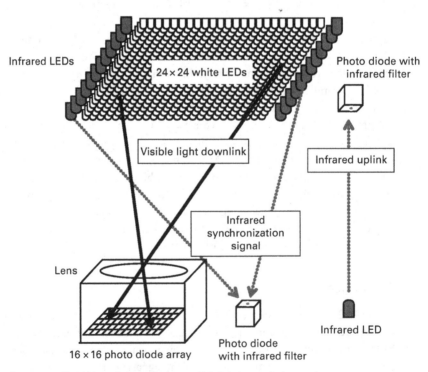

Figure 9.13   System configuration of a massively parallel data transmission system.

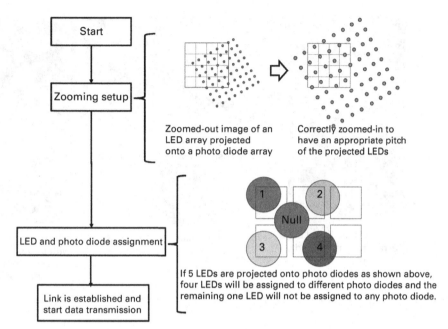

Figure 9.14   Link establishment sequence.

9.5.3   Link establishment

In parallel wireless VLC systems, multiple transmitted signals are spatially separated by a receiver lens. The image of the LED lighting transmitter is projected onto the PD array. All signals transmitted from different LEDs are independently received by each pixel of the image sensor. Before data transmission, we must first know the relationship between the transmitted signal and the receiving PD. Figure 9.14 shows the link establishment sequence.

The receiver first sets up the zoom of a lens by finding an appropriate pitch of the projected LEDs. Details of the zoom method can be found in [14]. Even if the zoom is correctly set up, cases with multiple LEDs projected onto the same PD will be present, as shown in Fig. 9.14, resulting in interference if different signals are sent from multiple LEDs to the same PD. Therefore, LEDs have to be correctly assigned to corresponding PDs to avoid such interference. This is achieved by sending appropriate information from receiver to transmitter using an uplink connection. The details of the LED and PD assignment can be found in [13]. The data transmission begins only after the completion of LED and PD assignment.

9.5.4   Prototype of a massively parallel data transmission system

Our actual prototype is shown in Fig. 9.15. The 24 × 24 visible light LED array transmitter is able to send individual data from each LED, each of which is controlled by the field-programmable gate array (FPGA) at the back of the LED board. At the receiver, the

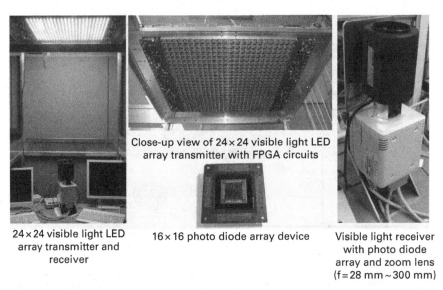

Close-up view of 24 × 24 visible light LED
array transmitter with FPGA circuits

24 × 24 visible light LED   16 × 16 photo diode array device   Visible light receiver
array transmitter and   with photo diode
receiver   array and zoom lens
                                                                                    (f = 28 mm ∼ 300 mm)

Figure 9.15   Prototype of a massively parallel data transmission system.

light from the transmitter goes through a zoom lens with a focal length ranging from
28 mm to 300 mm; the focused image is projected onto a 16 × 16 PD array device. The
two-dimensional PD array chip was made by Hamamatsu Photonics. The rate of the
transmitted signal from one LED is 5 Mbps; thus, when maximum parallelism is attained,
the theoretical maximum data rate is 16 × 16 × 5 Mbps = 1280 Mbps.

9.6   Accurate sensor pose estimation

9.6.1   Overview

In image sensor based VLC, it is possible to compute a sensor (i.e., camera) pose with
computer vision techniques [14–16]. This is one of the major advantages as compared
with other communication techniques. Below, we introduce the basis of computer vision
for pose estimation, as well as a pose estimation method combined with both computer
vision and VLC techniques.

9.6.2   Single view geometry

Computer vision is a field involving analyses of the geometry of the real world from the
images captured with a camera. In general, research issues are classified into the
estimation of the camera pose and the reconstruction of the three-dimensional shape of
an object (i.e., three-dimensional modeling) from an image or image sequence. Related
research fields of computer vision include image processing, pattern recognition, and
machine learning for scene recognition and understanding.

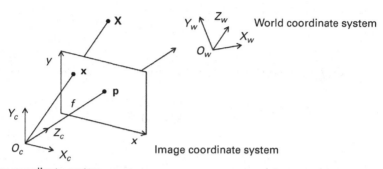

Figure 9.16   Single view geometry.

In image sensor based VLC, an image sensor can receive data simultaneously transmitted from multiple lights, because each pixel on the sensor is treated as a receiver. Further, the positions of the lights are acquired as feature points on the image sensor and can be used for camera pose estimation. Since this is a traditional research issue, substantial research is readily available in the literature on camera pose estimation using points [14]. Below, we introduce the basis of single-camera geometry for camera pose estimation. Note that the term "pose" here represents the position and orientation relative to some coordinate system in the field of computer vision.

As shown in Fig. 9.16, camera geometry is described via three coordinate systems, namely the three-dimensional world coordinate system, the two-dimensional image coordinate system, and the three-dimensional camera coordinate system. A camera pose is normally defined as the position and orientation of the camera coordinate system relative to the world coordinate system. The mathematical representation of the position and orientation is equivalent to the parameters of geometric transformation from the world coordinate system to the camera coordinate system as

$$\tilde{\mathbf{X}}_{\mathbf{c}} = \begin{bmatrix} \mathbf{R} & \mathbf{t} \\ 0 & 1 \end{bmatrix} \tilde{\mathbf{X}}_{\mathbf{w}},$$

where $\tilde{\mathbf{X}}_{\mathbf{c}} = (X_c, Y_c, Z_c, 1)^{\mathbf{T}}$ is a homogeneous camera coordinate system, $\tilde{\mathbf{X}}_{\mathbf{w}} = (X_w, Y_w, Z_w, 1)^{\mathbf{T}}$ is a homogeneous world coordinate system, $\mathbf{R}$ is a 3×3 rotation matrix (orientation), and $\mathbf{t}$ is a 3×1 translation vector (position). Therefore, a camera pose is equivalent to $[\mathbf{R}|\mathbf{t}]$.

The camera coordinate system is defined as the system with its origin located at the camera center and the direction of Z_c perpendicular to the image plane from the camera center. The intersection of the image plane and Z_c axis is called principal point $\mathbf{p} = (p_x, p_y)$. In the pinhole camera model, a three-dimensional point $\mathbf{X}_{\mathbf{c}} = (X_c, Y_c, Z_c)^{\mathbf{T}}$ in the camera coordinate system is projected onto a two-dimensional point $\mathbf{x} = (x, y)^{\mathbf{T}}$ in the image coordinate system as

$$(x,y)^{\mathrm{T}} = \left(f\frac{X_c}{Z_c} + p_x, f\frac{Y_c}{Z_c} + p_y \right)^{\mathrm{T}},$$

where f is the focal length of the lens. By establishing camera calibration matrix $\mathbf{A}$ as

$$\mathbf{A} = \begin{bmatrix} f & 0 & p_x \\ 0 & f & p_y \\ 0 & 0 & 1 \end{bmatrix},$$

the projection of a three-dimensional point in the world coordinate system onto a two-dimensional point in the image coordinate system is described as

$$\tilde{\mathbf{x}} \sim \mathbf{A}[\mathbf{R}|\mathbf{t}] \, \tilde{\mathbf{X}}_{\mathbf{w}},$$

where $\tilde{\mathbf{x}} = (x, y, 1)$ is a homogeneous image coordinate. This equation can be simplified as

$$\tilde{\mathbf{x}} \sim \mathbf{P}\tilde{\mathbf{X}}_{\mathbf{w}}$$

$$\mathbf{P} = \mathbf{A}[\mathbf{R}|\mathbf{t}],$$

where $\mathbf{P}$ is a 3×4 perspective projection matrix that also represents a camera pose.

To estimate a camera pose by solving the above equations, obtaining multiple sets of $\tilde{\mathbf{x}}$ and $\tilde{\mathbf{X}}_{\mathbf{w}}$ is mandatory. For example, $\mathbf{P}$ is linearly computed from six sets, because there are 12 unknown parameters in $\mathbf{P}$ and two equations are determined from one set. If $\mathbf{A}$ is known by using a camera calibration technique [17], it is possible to compute a camera pose with up to four pairs of solutions [18]. Because many solutions under different conditions have been proposed in the literature, solutions are not limited to the ones described above.

9.6.3   Pose estimation using lights

As explained above, acquiring multiple sets of a world coordinate and its projected image coordinate for camera pose estimation is necessary. In image sensor based visible light communication, such sets are acquired using lights.

Figure 9.17 shows an overview of pose estimation using lights. Lights are first placed in a target scene and their three-dimensional world coordinates are measured using an electronic distance meter, such as a total station. To compute the image coordinate of a light and receive its transmitted data, a camera captures the lights at a fixed position, because a light should be captured at the same position through an image sequence.

When transmitted data include the world coordinate of the light, we acquire the set containing the world coordinate of a light and its projected image coordinate; however, the quantity of transmitted data may not be enough. In such a case, the transmitted data can be just an identification number (ID), and the list of identification numbers of lights and their world coordinates should be stored, as shown in Table 9.2. If the number of acquired data sets satisfies the condition of pose estimation, a camera pose can be computed. Note that the

Table 9.2 IDs and world coordinates.

ID	World coordinate
1	(X_1, Y_1, Z_1)
2	(X_2, Y_2, Z_2)
$\vdots$	$\vdots$
N	(X_N, Y_N, Z_N)

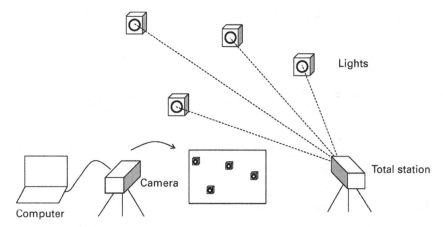

Figure 9.17   Pose estimation using lights.

accuracy of an estimated camera pose depends on the accuracy of the measurements of three-dimensional world coordinates. In the following subsection, we explain an efficient method for computing the image coordinate of a blinking light.

9.6.4   Light extraction

To extract lights in images, one approach is to use a predefined brightness threshold such that a pixel is judged as light if the measured brightness is more than the given threshold; however, this approach does not work well, because brightness is substantially affected by many elements, such as image sensors, environmental lights, the brightness of a light, and the distance between the camera and light. To stably extract lights, light extraction using the rules of blinking patterns was proposed in [17], and a flow chart is shown in Fig. 9.18 (please also refer to the figures in [17] for more illustrative details). Below, we explain each step of the light extraction process.

9.6.4.1   Computation of a threshold

First, a threshold for converting the brightness value into binary is adaptively computed at each pixel. Given an image sequence that includes N images, each pixel i has N brightness values; maximum brightness L_i and minimum brightness S_i are selected. Then, the new adjusted threshold at each pixel T_i is computed as

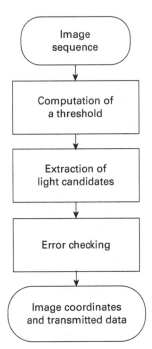

Figure 9.18   Flow chart of light extraction in [17, 19].

$$T_i = \frac{L_i - S_i}{2} + S_i.$$

9.6.4.2   Extraction of light candidates

After computing a binary sequence at each pixel using the threshold, one approach is to use an error-checking scheme embedded in the transmitted data to check all pixels; however, its computational cost may be huge; thus, decreasing the number of pixels becomes important for error checking. Therefore, to avoid the huge computational cost, light candidate pixels are extracted by checking whether a binary sequence follows the rules of the blinking pattern.

In [17], the blinking pattern is designed such that 1 bit is represented by 4 samples, i.e., 0011 is 0 and 1100 is 1. This means that 010 and 101 do not appear in the blinking pattern of three samples. If a pixel includes such a pattern, it can be removed. By checking the pattern of a binary sequence of several images, the number of pixels that do not capture light is drastically reduced.

9.6.4.3   Error checking

After extracting the light candidate pixels, a binary sequence of all images is computed for these pixels. Then a binary sequence is tested with an error-checking scheme embedded in the transmitted data to increase the reliability of light extraction. Next, the adjacent pixels that passed the error-checking phase and had the same binary sequence are connected to make a light area. Finally, by computing the center of each light area, the image coordinates of a light and the transmitted data are computed.

9.7   Applications of image sensor based communication

9.7.1   Traffic signal communication

In this subsection, we introduce a vehicle-to-infrastructure visible light communication (V2I-VLC) system using an LED array transmitter, which is assumed to be an LED traffic light, and an in-vehicle receiver equipped with a high-frame-rate (HFR) CMOS image sensor camera or high-speed camera (HSC) [4].

During our experiments, we placed an LED array horizontally on the ground and mounted the high-speed camera on the dashboard of the vehicle. The vehicle was driven directly toward the LED array at 30 km/h, as shown in Fig. 9.19. The communication distance in these field trials ranged from 110 to 30 m.

The transmitter consisted of 1024 LEDs arranged in a 32 × 32 square matrix. The LED spacing was 15 mm, and its half value angle was 26 degrees. To compensate for the vibrations of the car, we represented one data bit using four LEDs (a 2 × 2 LED array). Each LED blinks at 500 Hz, whereas the PWM was processed at 4 kHz. We used R = 1/2 turbo code for error correction and inverted LED patterns for tracking. Accordingly, the overall data rate was 32 kbit/s (= 500 bit/s × 256 × 1/2 × 1/2). The input data were audio data and were assumed to be transmitting safety information transmitted from LED traffic lights.

For the receiver, we used an in-vehicle HSC (i.e., a Photoron FASTCAM 1024PCI 100k) with a frame rate of 1000 fps and a resolution of 512 × 1024 pixels connected to a personal computer (PC). The focal length of the lens was 35 mm. In general, the light sensitivity of high-speed image sensors was set high to provide rapid exposure times, which also means we could set a relatively small lens aperture. For example, the ISO sensitivity of the HSC was set at 10 000, and the lens aperture was set to 11. In addition, since autofocusing is difficult when a vehicle is moving, the focus was set to infinity.

Figure 9.19   Experimental equipment.

The header image-processing, the data image-processing, and the decoding tasks were performed by the PC. We also recorded and displayed a gray scale video, obtained using the HSC as a drive recorder, which simultaneously recorded the view in front of the vehicle and the data transmitted from the LED array. We confirmed the robust detection and tracking of the LED array with respect to camera vibration, along with a lack of error in LED array detection and tracking. Next, we confirmed a clear audio signal (32 kbit/s) reception for distances of up to 45 m with error-free performance.

We also conducted an experiment on the simultaneous transmission of text information. In this case, the data rate was 2 kbit/s and error-free performance was achieved from 110–20 m, which is deemed to be a suitable range for intersection safety applications.

9.7.2   Position measurements for civil engineering

As introduced in Section 9.6 above, a camera pose can be computed in the framework of single view geometry. If multiple cameras can be used, the three-dimensional position of an object in the real world can be computed [20]. In this subsection, we introduce triangulation with image sensor based VLC and its application to civil engineering [14].

9.7.2.1   Basis of triangulation

Triangulation is a method for three-dimensional position measurement of an object in the real world with two known cameras. Because triangulation is closely related to camera pose estimation, we first briefly revisit perspective projection introduced in Section 9.6.

A three-dimensional point in the world coordinate system is projected onto a two-dimensional point in the image coordinate system. When three-dimensional point $\widetilde{X}_w$ is projected onto two cameras, this is mathematically described as

$$\widetilde{x}_1 \sim P_1 \widetilde{X}_w$$

$$\widetilde{x}_2 \sim P_2 \widetilde{X}_w,$$

where P_i is a perspective projection matrix of each camera and $\widetilde{x}_i$ is a homogeneous image coordinate computed by projecting $\widetilde{X}_w$ onto each image plane with P_i; $\widetilde{x}_1$ and $\widetilde{x}_2$ are considered to correspond between two images. Use of two known cameras means that P_1 and P_2 are known. This is also graphically explained in Fig. 9.20. In triangulation, the goal is to compute $\widetilde{X}_w$ by finding $\widetilde{x}_1$ and $\widetilde{x}_2$.

9.7.2.2   Triangulation with image sensor based VLC

If we put a blinking light at $\widetilde{X}_w$, it is easy to find $\widetilde{x}_1$ and $\widetilde{x}_2$ with image sensor based VLC. In each camera, the image coordinates of lights and the transmitted data are first computed as introduced in Section 9.6 above. Then $\widetilde{x}_1$ and $\widetilde{x}_2$ can be acquired by selecting the pixels that receive the same data in each camera.

9.7.2.3   Application to bridge shape monitoring

Image sensor based VLC works under different types of illumination from daybreak to midnight, because a light can actively transmit data by blinking. This feature is very useful

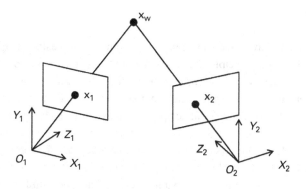

Figure 9.20   Two view geometry.

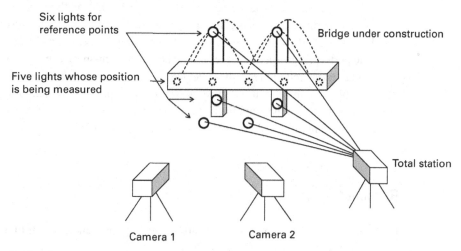

Figure 9.21   Bridge shape measurement.

for bridge shape monitoring. In general, the shape of a bridge is deformed when the fabric of the bridge is dilated and shrinks according to the ambient temperature. When a bridge is built, such deformation should be monitored for the early detection of problems.

Figure 9.21 shows an overview of bridge shape monitoring using an image sensor based VLC approach. On the bridge, two different types of lights are placed, i.e., points for measuring the deformation and points for camera pose estimation. The latter points are typically called reference points and should be placed at points where deformation does not occur.

Then, the monitoring procedure is as follows. From the reference points, the poses of camera 1 and camera 2 are first computed. Then, in each camera, the image coordinates of the lights and the transmitted data are computed. By selecting the pixels that receive the same data, the corresponding pixels of the light in the two images are computed. These pixels are triangulated to compute the three-dimensional position of the light.

9.8　Summary

In this chapter, we have covered visible light communication using image sensors, starting with its fundamental principles, followed by application examples of massively parallel communication, accurate sensor pose estimation, traffic signal communication, and position measurement for civil engineering. An image sensor is able to spatially separate visible light sources, receive and demodulate optical signals from visible light transmitters at pixel positions where optical light is projected, and detect accurate arrival angles of incoming light. By using the unique characteristics of the image sensors, the applications described in this chapter are made possible. For example, the use of multiple pixels as receivers makes it possible to achieve massively parallel communication; further, the capability of very accurate arrival angle detection makes it possible to perform accurate sensor pose estimation and position measurement for civil engineering; finally, the capability of image processing makes it possible to achieve traffic signal communication.

Many of the examples described above are for special applications and are still somewhat experimental. This is primarily due to the high cost of image sensor receivers and the technical difficulty of high-speed data reception. Conventional image sensors are unable to perform high-speed data reception; however, high-speed cameras could be used if high data rates are required. These devices are currently prohibitively expensive and not available for consumer use, but when useful VLC applications using image sensors are developed – and if these applications are widely adopted – the cost will go down and make these techniques easily available in consumer applications.

References

[1] S. Haruyama and T. Yamazato, "[Tutorial] Visible light communications," IEEE International Conference on *Communications*, Jun. 2011.

[2] Stuart A. Taylor, "CCD and CMOS imaging array technologies: Technology review," Technical Report EPC-1998–106, 1998.

[3] Dave Litwiller, "CCD vs. CMOS: Facts and fiction," *Photonics Spectra*, Jan., 2001.

[4] Takaya Yamazato, Isamu Takai, Hiraku Okada, *et al.*, "Image sensor based visible light communication for automotive applications," *IEEE Communication Magazine*, **52**, (7), 88–97, 2014.

[5] Photoron FASTCAM SA-X2, http://www.photron.com/?cmd=product_general%product_id=39

[6] T. Nagura, T. Yamazato, M. Katayama, *et al.*, "Improved decoding methods of visible light communication system for ITS using LED array and high-speed camera," IEEE *Vehicular Technology* Conference (VTC-Spring2010), May 2010.

[7] H. B. C. Wook, S. Haruyama, and M. Nakagawa, "Visible light communication with LED traffic lights using 2-dimensional image sensor," *IEICE Trans. on Fundamentals*, **E89-A**, (3), 654–659, 2006.

[8] S. Nishimoto, T. Yamazato, H. Okada, *et al.*, "High-speed transmission of overlay coding for road-to-vehicle visible light communication using LED array and high-speed camera," IEEE Workshop on *Optical Wireless Communications*, pp. 1234–1238, Dec. 2012.

[9] S. Arai, S. Mase, T. Yamazato, *et al.*, "Feasibility study of road-to-vehicle communication system using LED array and high-speed camera," Proceedings of the 15th World Congress on ITS, Nov. 2008.

[10] Zabih Ghassemlooy, Wasiu Popoola, and Sujan Rajbhandari, *Optical Wireless Communications: System and Channel Modelling with MATLAB®*, CRC Press, 2012.

[11] G. Yun and M. Kavehrad, "Indoor infrared wireless communications using spot diffusing and fly-eye receivers," *Canadian Journal of Electrical and Computer Engineering*, **18**, (*4*), 151–157, 1993.

[12] Joseph M. Kahn, Roy You, Pouyan Djahani, *et al.*, "Imaging diversity receivers for high-speed infrared wireless communication," *IEEE Communications Magazine*, **36**, (*12*), 88–94, 1998.

[13] Satoshi Miyauchi, Toshihiko Komine, Teruyuki Ushiro, *et al.*, "Parallel wireless optical communication using high speed CMOS image sensor," International Symposium on *Information Theory and its Applications (ISITA)*, Parma, Italy, 2004.

[14] Masanori Ishida, Satoshi Miyauchi, Toshihiko Komine, Shinichiro Haruyama, and Masao Nakagawa, "An architecture for high-speed parallel wireless visible light communications system using 2D image sensor and LED transmitter," in Proceedings of International Symposium on *Wireless Personal Multimedia Communications*, pp. 1523–1527, 2005.

[15] T. Yamazato and S. Haruyama, "Image sensor based visible light communication and its application to pose, position, and range estimations," *IEICE Trans. on Commun.*, **E97.B**, (*9*), 1759–1765, 2014.

[16] Richard Szeliski, *Computer Vision: Algorithm and Applications*, Springer, 2011.

[17] Zhengyou Zhang, "A flexible new technique for camera calibration," *IEEE Transactions on Pattern Analysis and Machine Intelligence*, 1330–1334, 2000.

[18] David Nister, "A minimal solution to the generalised 3-point pose problem," in Proceedings of the IEEE Computer Society Conference on *Computer Vision and Pattern Recognition*, pp. 560–567, 2004.

[19] Hideaki Uchiyama, Masaki Yoshino, Hideo Saito, *et al.*, "Photogrammetric system using visible light communication," in Proceedings of the 34th Annual Conference of the IEEE Industrial Electronics Society, pp. 1771–1776, 2008.

[20] Richard Hartley and Andrew Zisserman, *Multiple View Geometry in Computer Vision*, Cambridge University Press, 2004.

Index

Figures and tables are denoted in bold typeface.